# Short Essays
## for
# Inquiring Minds

Ronald Gruner

**SHORT ESSAYS FOR INQUIRING MINDS**
2025 Edition
Copyright © 2026 Ronald Gruner. All rights reserved.

No part of this book may be reproduced, stored in a retrieval system, or transmitted by any means, electronic, mechanical, photocopying, recording, or otherwise, without written permission from the copyright holder.

Published by Libratum Press
Naples, Florida

Cover: Herbert J. Ona
Index: Sergey Lobachev

Softcover ISBN: 978-1-7378231-9-3
eBook ISBN: 979-8-9944392-0-3
Library of Congress Control Number: 2026900005

*For*
*Lindsay, Julie, and Nicole*

# Contents

**INTRODUCTION** ........................... 1

## COVID & PUBLIC HEALTH
A Mysterious Malady in Tobacco Plants
    Led to the Discovery of Viruses ..................................................... 3
A Lethal New Virus ............................................................................. 15
The Beginning .................................................................................... 19
President Trump and Dr. Fauci Part Ways ........................................ 33
Alternative Facts ................................................................................ 39
Dr. Atlas and America's Giant Measles Party ................................... 46
Vaccines: a Mighty and Horrible Monster ........................................ 59
The Great Barrington Declaration ..................................................... 64

## PRESIDENTIAL LEADERSHIP
The Warning ........................................................................................ 9
Nine Millionaires and a Plumber ...................................................... 26
Harding, Trump, and America First .................................................. 52
President Trump Declares COVID a National Emergency ............... 71
How President Clinton Let Russia Slip Away ................................... 73
The Life and Times of William McKinley ......................................... 97
The Corpse at Every Funeral, the Bride at Every Wedding ............ 110
Measuring President Trump's First 100 Days in Office .................. 115
President Washington's Greatest Disappointment ........................ 173
President Trump Announces New Tariffs
    and Fires Dr. Erika McEntarfer ............................................... 206
Letters from the Past: Four Presidents on Today's America ........ 325

## ECONOMIC POLICY

A Trade War over Chickens
    Led to America's Love for Pickup Trucks..................................86
Looted, Pillaged, Raped, and Plundered..........................................92
Tariffs, Gold, Silver, and the Wizard..............................................104
President Trump and Pope Leo XIV ..............................................124
On Workers and Capital..................................................................129
How to Make America Great Again................................................134
It's Déjà Vu All Over Again............................................................140

## FOREIGN POLICY

United States and Iranian Relations: 1953–1988..........................160
Greetings from Greenland..............................................................193
The Berlin Airlift ............................................................................211
Don't Cry for Me Argentina...........................................................283

## DEMOCRACY UNDER PRESSURE

The Rise and Fall of the Department of Education.........................79
The Surprising Origins of Social Security .....................................145
Is The Republican Pledge Obsolete?..............................................216
"Gerrymandering is Incompatible
    with Democratic Principles."...................................................222
A Defiant RFK Jr., Poor Jobs Report,
    and New Department of War ..................................................235
Congress Shuts Down the Government ........................................258
Just How Serious is the Federal Deficit? ......................................264
America's Neglected Air Traffic Control System..........................288
Immigration: Culture versus Economics .......................................295
The Danger of Armed Government Intervention .........................306
Immigration and Violent Crime.....................................................312

## ARTIFICIAL INTELLIGENCE

ChatGPT Analyzes Joe Biden and Donald Trump .......................152
Will AI be Writing Newspaper Editorials Soon?...........................167

A Brief History of Artificial Intelligence: Part 1 ............................ 180
A Brief History of Artificial Intelligence: Part 2 ........................... 186
A Brief History of Artificial Intelligence: Part 3 ............................ 271
Can Artificial Intelligence Repair America's Media Divide? ......... 277
A Brief History of Artificial Intelligence: Part 4 ............................ 301

## MEDIA & CULTURE
Cracker Barrel Gives Uncle Herschel the Boot ............................ 229
Charlie Kirk and Martin Luther King Jr ....................................... 242
Charlie Kirk in His Own Words .................................................... 247
The Boston Crusaders .................................................................. 254
Money and the Media .................................................................. 319

## LAST BUT HARDLY LEAST
The Short, Miserable Life of a Golf Ball ....................................... 331

## ABOUT THE AUTHOR ................ 335

## INDEX ........................................... 337

# Introduction

The fifty-four essays included in this book are a compilation of the weekly articles I posted on my Substack website from December 2024 through the end of 2025.

Substack is a digital publishing platform that allows writers to develop their own media channels offering a direct connection with subscribers at no cost to either the publisher or subscriber. In short, Substack is social media but, at least today, responsible social media with little disinformation, insults, or abusive language.

I have always enjoyed writing starting on the high school newspaper. During my career, I spent forty-five years in technology, first as an engineer and later as an entrepreneur. During those years I often wrote, everything from technical papers to news releases. After retiring, I began to write for myself.

My first book, *We the Presidents*, was published in 2022. The book explores how American presidents over the last century from Warren G. Harding to Donald J. Trump have shaped America and the world. Three years later, I published *Covid Wars* which discusses the tension between public health and personal freedom during the Covid pandemic.

Writing a book is a long, lonely task often taking years. That's why I started my weekly Substack. Every week offers a fresh topic, typically 1,200 words requiring just a day or two to write, and five or six minutes to read. And rather than waiting years for a book to be published, the response from readers arrives weekly.

I started writing these short essays in late 2024, shortly after Donald Trump won the presidential election. I had written extensively about his first presidential term in *We the Presidents* and knew the next four years would provide plenty of engaging material. That certainly has been the case.

But the world is much more interesting than Washington politics. So, I've tried to engage readers with diverse topics from artificial intelligence to Pope Leo XIV. That seems to have worked. Three of the most popular essays over the last fifteen months were very different: "Will AI be Writing Newspaper Editorials Soon?" "Charlie Kirk in His Own Words," and "America's Neglected Air Traffic Control System."

In writing, I've generally tried to follow Eleanor Roosevelt's rather pretentious declaration: "Great minds discuss ideas; average minds discuss events; small minds discuss people." So, I've focused on ideas surrounding economics, technology, and public health. But it's events, and even more so, people, that make life interesting. So, I've also written about topics such as diverse as the Berlin Airlift, the life of an Uber driver, and Cracker Barrel's aborted logo change.

In writing these essays, I've learned that it's tempting to grow readership by focusing on topics which reinforce readers' existing beliefs and biases. In short, telling an audience what they want to hear, often at the expense of truth. Too much of today's social and cable media, focused on subscription and advertising revenue, have done that. I've tried not to do so. This book is priced only slightly above break-even, and as I promise my readers, a subscription to *RonaldGruner.Substack.com* is "free and always will be."

Ronald Gruner
January 2026

# A Mysterious Malady in Tobacco Plants Led to the Discovery of Viruses.

DECEMBER 10, 2024

Research into the cause of a mysterious malady in tobacco plants led to the discovery of viruses. Starting in 1885, tobacco plants around the world had begun to fail after their leaves turned a mottled yellow. The disease spread easily among the plants, reducing crop yields, threatening tobacco companies and even the supply of cigarettes to the Russian Army.

Scientists suspected that the tobacco disease was caused by some type of bacteria. Twenty-five years earlier, the French chemist, Louis Pasteur, had developed the germ theory of disease. For centuries, physicians had believed in "spontaneous generation," that the pathogens responsible for disease—named bacteria in 1828—arose spontaneously from decaying organic matter. But Pasteur, through a series of simple experiments, demonstrated that bacteria spread from external sources rather than emerging spontaneously.

In 1890, Russian Czar Alexander III ordered Dmitri Ivanovsky, a Russian botanist, to research the disease killing tobacco plants in Crimea. Two years later, Ivanovsky attempted to isolate the responsible bacteria by filtering the sap of diseased tobacco plants through a Pasteur-Chamberland filter, an extremely fine ceramic filter which would block any known bacteria. (Ceramics are porous. It's the glaze on the outside which keeps your coffee cup from leaking.)

Yet when Ivanovsky exposed tobacco plants to the filtered sap, they somehow became infected. Ivanovsky concluded that an extremely small infectious agent, too small to be seen under a microscope, must be passing through the filter. Even more confounding, while the filtered sap was infectious to plants, Ivanovsky was unable to culture the infectious agent in a laboratory. The mysterious new pathogen, unlike bacteria, needed to invade a living cell to replicate.

In 1892, Dmitry Ivanovsky was the first to isolate a virus.

In 1898, Dutch microbiologist Martinus Beijerinck independently confirmed that the infectious agent in the plant sap passed through the finest ceramic filters available. Beijerinck publicized his findings in a historic paper titled, "Contagium Vivum Fluidum." Unable to accept that a sub-microscopic bacteria could exist, Beijerinck claimed the infectious agent was a "contagious living fluid." Beijerinck called the mysterious agent a "virus"; being the Latin word for poison, it was a well-chosen name.

Ivanovsky and Beijerinck had discovered a fundamentally new pathogen lying somewhere between an inert molecule and a living cell. They were the first virologists.

Scientists soon discovered other diseases were caused by the invisible virus, including cattle foot-and-mouth disease and yellow fever. One virus killed more than twenty million people.

## The 1918 Spanish Flu

The 1918 Spanish Flu, also known today as the H1N1 influenza, broke out in early 1918 as the First World War was raging in Europe. Killing over twenty million people during its two-year rampage, the Spanish Flu was the deadliest pandemic since the Black Death pandemic during the fourteenth century.

The first identified case occurred on March 11, 1918, at Fort Riley, Kansas. The new virus was highly contagious, and within a week over five hundred soldiers had been admitted to the camp hospital suffering flu-like symptoms. Weeks later the virus had spread to the East Coast as soldiers embarked on their voyage to Europe in cramped troop ships.

By summer, the contagion had spread throughout Europe, killing millions and decimating the warring armies. Fearful of revealing their weakened conditions to the enemy, the warring nations censored reports of their influenza deaths, except for Spain. Still recovering from its loss of Cuba and the Philippines during the Spanish-American War of 1898, Spain remained neutral. Spain's uncensored newspapers regularly reported the country's skyrocketing death count, creating the false impression that Spain was the pandemic's epicenter. The Spanish Flu misnomer, naturally, followed.

Influenza kills those whose immune systems cannot fight off the virus. That's usually the very young and the elderly. But the Spanish Flu killed healthy, young adults, leaving their orphaned children behind. Even today, virologists are not sure why although they suspect the virus triggered an overreaction by the immune system known as a cytokine storm. Death often came in hours, starting with a sore throat, cough, and fever and progressing to difficulty breathing as the victim's lungs filled with fluid, then ending in a suffocating death as the victim slowly drowned in their own fluids.

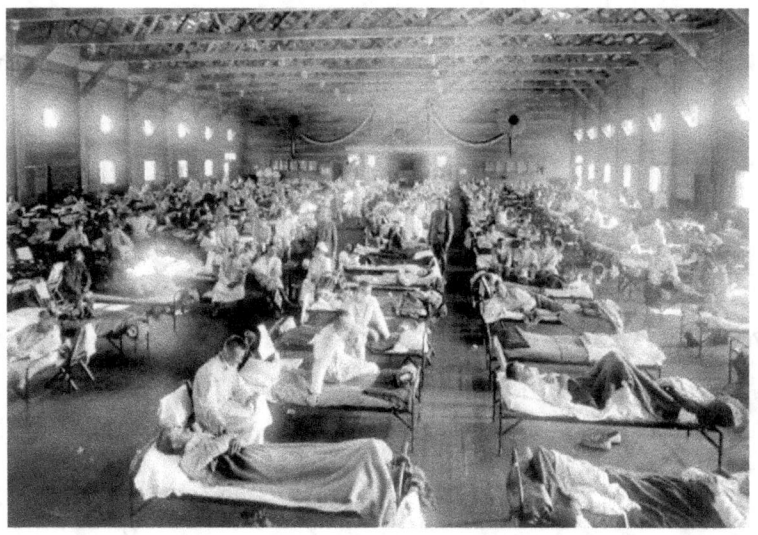

The influenza ward at Camp Funston, Kansas in the spring of 1918

The second and third influenza waves hit the United States especially hard in the fall and winter of 1918. It was a horrible time. Across the country, Americans were divided on how to respond to the virus. Two cities, Philadelphia and St. Louis, encapsulated the differences.

On September 28, 1918, Philadelphia held a patriotic Liberty Bond parade—an important fund-raiser during the war—which was attended by an estimated 200,000 people. The parade organizers had ignored the U.S. Surgeon General's warning just a few days earlier that "influenza is a crowd disease" and that crowds should be avoided.

The parade route stretched four miles through the center of the city. Parade marchers mingled with the crowds, eating and drinking as the parade progressed and making it the perfect incubator for the spread of the virus. Within days, influenza deaths began to rise. By the end of October, 11,000 Philadelphians had died. With the hospitals filled to capacity, many died in their homes. Deaths were so numerous that the city's streetcars were used as hearses to carry the dead directly to overwhelmed cemeteries.

St. Louis responded differently. The city cancelled its Liberty Bond parade when the city's respected health official, Dr. Max Starkloff, advised against it in spite of heavy pressure from the business community. Starkloff even ordered stores closed and celebrations muted on Armistice Day, November 11, recognizing the end of the First World War.

The difference between the two cities was stark. From September through December, Philadelphia suffered 13,936 influenza deaths while St. Louis had only 2,883. In spite of the risks, and motivated by the call to wartime patriotism, people resisted public health measures. Public health officials struggled to convince citizens to accept mitigation measures such as school, church, and business closures. Face masks were unpopular and abused, often humorously; men would cut holes in their masks so they could smoke cigars.

By April 1919, the pandemic had largely ended. In a little more than a year, an estimated 675,000 Americans had died from the virus—far more than the 117,000 American deaths during the First World War.

Today, epidemiologists believe the Spanish Flu virus originated in birds and then jumped to humans through an intermediate host, possibly swine. But during the pandemic, doctors mistakenly believed the disease was bacteria-based. It wouldn't be until 1933 that British researchers at London's National Institute for Medical Research isolated and identified the 1918 influenza virus. Distant descendants of the 1918 influenza virus continue to circulate throughout the human population, making the 1918 virus "truly the 'mother' of all subsequent influenza pandemics."

# The Warning

DECEMBER 17, 2024

During August 2005, much of Washington was chuckling over President Bush's summer reading list recently distributed by the White House. Like his father, who during his 1991 vacation commented that he "[played] a good deal of tennis, a good deal of horseshoes, a good deal of fishing, a good deal of running—and some reading," the younger Bush was known for vacations spent fishing and clearing brush at his Texas ranch.

But the White House reading list suggested the younger Bush was a more serious reader than his cowboy image suggested: *Salt: A World History* by Mark Kurlansky, *Alexander II: The Last Great Tsar* by Edvard Radzinsky, and *The Great Influenza: The Story of the Deadliest Pandemic in History* by John Barry. Hardly light summer reading, the president's critics couldn't imagine Bush plodding through the three books' combined 1,500 pages. But the president's reading list was well chosen.

As a former oilman, Bush was struck by the analogies between salt and oil. Four centuries ago, Queen Elizabeth I warned that Britain had become too dependent on foreign salt, a strategic material essential for food preservation. Since the 1974 oil embargo, American presidents have had the same concerns regarding foreign oil.

Bush also knew that President Lincoln and Russian Czar Alexander II had a warm rapport, exchanging five letters during the Civil War. Both leaders grappled with slavery. Alexander II freed twenty million Russian serfs when he issued his Edict of Emancipation in 1861. Two years later, Lincoln's Emancipation Proclamation freed the slaves in the eleven Confederate states. Both men were later assassinated for their reforms.

But it was John Barry's book, *The Great Influenza*, that most influenced Bush. The book examines how the 1918 Spanish Flu became the deadliest killer in centuries after starting as a small outbreak in Kansas. Within months the disease erupted into a worldwide pandemic that, ultimately, killed over twenty million people. The world, exhausted from the First World War and with few medical weapons to combat the virus, was unprepared.

Fourteen years before the COVID virus emerged, President George W. Bush proposed legislation to help manage a major pandemic. His warning was soon forgotten.

After reading Barry's book, Bush was determined that the United States would be prepared when the next pandemic struck the country, as it inevitably would. The September 11 terrorist attacks had killed three thousand Americans; a pandemic could kill millions. Two months after returning from vacation, on November 1, 2005, President Bush gave a major policy speech on how the federal government should prepare for an influenza pandemic. The president set three goals.

The first was to quickly detect outbreaks before they spread. "A pandemic is a lot like a forest fire," Bush said. "If caught early it might be extinguished with limited damage. If allowed to smolder, undetected, it can grow to an inferno that can spread quickly beyond our ability to control it."

The second goal was to accelerate vaccine development, especially vaccines based on cell-culture technology. "By bringing cell-culture technology from the research laboratory into the production line," Bush said, "we should be able to produce enough vaccine for every American within six months of the start of a pandemic."

The third goal was to establish clear emergency plans so the nation would be prepared to respond quickly and unequivocally when a pandemic threatened.

The president made another request, one that is often condemned today: Bush proposed that the federal government shield drug companies from litigation related to vaccines. In his November 1 speech, Bush asserted:

> "In the past three decades, the number of vaccine manufacturers in America has plummeted, as the industry has been flooded with lawsuits. Today, there is only one manufacturer in the United States that can produce influenza vaccine. That leaves our nation vulnerable in the event of a pandemic. We must increase the number of vaccine manufacturers in our country and improve our domestic production capacity. So, Congress must pass liability protection for the makers of life-saving vaccines."

President Bush's proposal to shield vaccine manufactures from litigation was, and continues to be, controversial for understandable reasons. In March 2024, for example, U.S. Representative Chip Roy introduced legislation to make vaccine manufacturers liable for injuries resulting from their COVID vaccines. "Millions of Americans were forced to take a COVID-19 shot..." Representative Roy commented in a prepared statement. "Many have faced injury from the vaccine, but few have been afforded little recourse.... The American people deserve justice for the infringement on their personal medical freedom and those medically harmed deserve restitution."

Representative Roy's proposed legislation quickly died in committee, but shouldn't vaccine manufacturers be accountable for their mistakes? Certainly, as a Republican president, Bush was highly sensitive to the dangers of government intrusion into private markets. Yet Bush understood that few companies, accountable to their shareholders, would invest billions for the development of a vaccine that may never be needed, and if used, might bankrupt the company by lawsuits.

Shortly after the president's speech, the Department of Health and Human Services released a 396-page pandemic study citing the importance of rapid federal and state coordination during a pandemic. The study estimated that during an influenza pandemic as many as ninety million Americans would become infected and nearly two million would die.

To fund his ambitious plan, President Bush asked Congress to appropriate $7.1 billion in emergency funding. After working its way through Congress, Bush signed the Pandemic and All-Hazards Preparedness Act (Preparedness Act) on December 19, 2006.

Bush's support for pandemic planning was visionary, helped by memories of how unprepared America had been leading up to the 9/11 terrorist attacks. But budget pressures, politics, and apathy would slowly erode the Preparedness Act. By early 2020, Bush's visionary plan had been reduced to the point that President Trump could complain, "We took over an empty shelf. We took over a very depleted place, in a lot of ways."

But in 2009, the United States was ready. That year, the Preparedness Act met its first test.

In April 2009, a new strain of influenza virus emerged in a small mountain village in Veracruz, Mexico. Five-year-old Edgar Hernandez became patient zero when he contracted a severe, flu-like illness that doctors identified as resulting from a new influenza strain. They named it Swine Flu after Edgar's mother blamed the virus on a huge pig farm nearby.

Five-year-old Edgar Hernandez was "patient zero" during the 2009 Swine Flu Pandemic.

The Swine Flu virus is a descendent of the H1N1 Spanish Flu virus which, even after a century, continues to circulate in the human population. The new virus, though, consists of an unusual combination of human, avian, and swine genetic elements, which raised concerns that much of the world's population would have only limited natural immunity.

The virus spread rapidly, and on June 11 the World Health Organization declared the Swine Flu a global pandemic. Like the Spanish Flu a century earlier, the Swine Flu largely killed children and young adults with less developed immune systems.

Remarkably, by October researchers had developed a new vaccine based on traditional egg-based cultivation in which the virus is cultivated in fertilized chicken eggs, harvested, and then inactivated to create the vaccine. The World Health Organization organized a worldwide vaccination campaign that met little resistance. Local public health authorities coordinated the distribution of masks, school and event closings, and other measures to slow the spread of the virus.

The coordinated, worldwide response to the virus worked. By late 2009 the initial wave had subsided, and on August 10, 2010, the World Health Organization declared an end to the global pandemic.

More than 150,000 people worldwide died of the Swine Flu during 2009 and 2010 including 12,500 in the United States. Little Edgar Hernandez survived, though, cheerfully attributing his recovery to ice cream.

# A Lethal New Virus

JANUARY 7, 2025

The precursor of the virus responsible for the COVID-19 pandemic was discovered in 1965, isolated from the nasal swabbing of a child with a common cold. Unlike earlier, virologists could now use powerful, new electron microscopes to study the tiny pathogens.

What they saw was remarkable. The new virus was roughly spherical with numerous club-like protrusions extending outward from its surface. Some saw the protrusions as similar to solar corona, the hot gases ejected from the surface of the sun. Others likened the protrusions to the spikes adorning the top of crowns. A more accurate analogy, though, would be a medieval battering ram. But rather than attacking castles, the virus's spikes were used to weaken the walls of healthy cells, allowing the virus to invade the cell and replicate itself.

The new strain of viruses was named coronaviruses, corona being the Latin word for crown. Seven coronaviruses have, to date, been discovered. For nearly forty years coronaviruses were thought primarily to infect animals. That changed in November 2002 when a deadly coronavirus, SARS-CoV-1 (Severe Acute Respiratory Syndrome CoronaVirus), emerged in the Guangdong Province of China.

Travelers soon spread the virus. One super-spreader from Guangdong infected sixteen fellow travelers staying at Hong Kong's Metropole Hotel. The travelers then spread the virus to Canada, Singapore, Taiwan, and Vietnam. SARS, though, was not typically contagious, spreading mostly through close contact in hospitals, restaurants, and hotels. Unlike the COVID-19 outbreak twenty years later, SARS was most contagious when the carrier was symptomatic and its victims typically isolated.

Over the next eight months, the World Health Organization estimated that 8,096 people contracted the disease, 774 of whom died representing an overall death rate of 9.6 percent. Few aged twenty-four years and younger died, less than 1.0 percent. But for those over sixty-five, contracting SARS was nearly a death sentence: 55 percent died once infected.

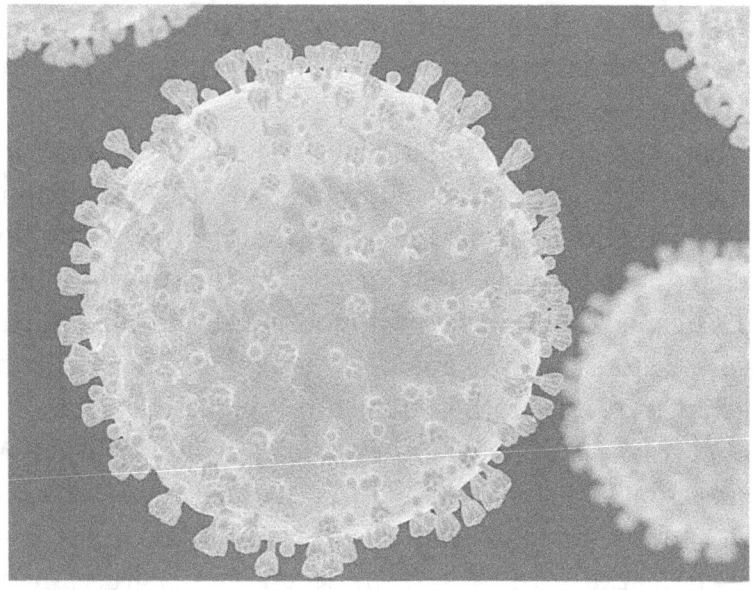

Computer-generated image of a coronavirus as seen through an electron microscope.  WIKIMEDIA

It would be fifteen years before scientists located the source of the virus that triggered the SARS outbreak. In 2017, researchers from the Wuhan Institute of Virology found a remote cave in Yunnan province that was home to horseshoe bats. The bats carried a coronavirus strain nearly identical to that which triggered the SARS outbreak. Rather than passing the virus directly to humans, the bats passed the virus on to wild animals, notably civet cats and pangolins.

A civet cat is a small, mostly nocturnal mammal dwelling in tropical forests. Pangolins, often called scaled anteaters, are solitary mammals that live in hollow trees and burrows. The animals, caught in the wild, were then sold in "wet markets," which specialized in exotic, fresh meat. Interestingly, the secretions from the civet cat's perineal gland near the base of their tail were for decades a critical ingredient in premium perfumes including Chanel #5.

The consumption of wild animals is common in China and Southeast Asia. It's been estimated that the Chinese market for wild animals is larger than the American beef industry. Snakes, civet cats, and pangolins are particularly popular. According to the British medical journal, *The Lancet*:

> "The civet cat has long been considered a delicacy, valued for its 'nutritious' meat, particularly in the winter months in southern China and Vietnam, with some tourists travelling to those regions specifically to eat civets and other exotic animals…. A survey done in Hong Kong in the late nineteen-nineties found that over 50 percent of respondents had eaten wild animals with the most common animal consumed being snakes, followed by civet cats (30 percent) and pangolins."

The next coronavirus outbreak occurred a world away from the humid jungles of Southeast Asia. In June 2012, the first case of MERS-CoV (Middle East Respiratory Syndrome CoronaVirus) was discovered in Jeddah, Saudi Arabia. The virus was spread primarily by dromedary camels rather than human contact. The Saudi Ministry of Agriculture issued an advisory to stay away from camels or wear a face mask when around them.

Many refused to comply. No animal is more beloved by Arabs than the camel. With their long eyelashes, smiling mouth, and graceful stride, camels are known as the "swan of the desert." Rather than shun the animals, many fondly kissed their camels on social media in defiance of the Ministry's order.

It is not known how camels acquired the virus, although bats, immune to the virus they're harboring, are suspected to have been the original carrier.

Unlike the SARS outbreak which lasted less than a year, MERS has been persistent, and much more lethal. Although human-to-human transmission is rare, it often occurs in health care settings where the MERS virus can spread to vulnerable individuals. From 2012 through 2022, the World Health Organization reported only 2,600 laboratory-confirmed cases of MERS with 84 percent of the cases confined to Saudi Arabia. During that period, 935 died of the disease, a staggering death rate of 36 percent.

The SARS and MERS outbreaks together killed several thousand people. A tragedy at the time, but viruses were constantly evolving as they had for millennia. By 2019, an airborne virus had evolved (or perhaps been created) that spread asymptomatically, silently infecting its unsuspecting victims. Its name was COVID.

Over the next five years COVID, the most lethal virus since the 1918 Spanish Flu, would claim more than seven million lives.

# The Beginning

JANUARY 14, 2025

In late December 2019, the first hints of an impending pandemic began to trickle out of China. They were spotted not by the World Health Organization (WHO) or the Centers for Disease Control and Prevention (CDC), but by Avi Schiffmann, a bright seventeen-year-old working from his bedroom in Mercer Island, Washington. Surfing the internet over the Christmas holidays, Avi had noticed a spike in influenza-like deaths in Wuhan, China.

Frustrated by the lack of information and already a proficient coder, Avi built a website to track the spread of the new disease. The website went live on December 31 while Avi and his family were on a weekend ski vacation. (Today, Avi's COVID tracking website remains one of the best.)

That same day, Chinese health authorities informed WHO that a new virus, similar to SARS, had been observed in Wuhan. The new virus strain was officially named Severe Acute Respiratory Syndrome Coronavirus 2 (SARS-CoV-2, or simply COVID) on February 11 after it was confirmed to be related to the 2002 SARS virus. At the same time, the resulting disease was named CoronaVirus Disease 2019 (COVID-19).

International travelers quickly spread COVID around the world. On February 1, the Philippines reported the first death outside China. Three days later, Hong Kong reported two deaths. Japan reported its first death on February 8, followed by

France on February 15, Taiwan on February 16, Iran on February 19, South Korea on February 20, and Italy a day later.

Avi Schiffmann was a high school junior when he created the first COVID tracking website in December 2019.

On March 11, the World Health Organization declared the COVID-19 outbreak to be a world pandemic. By then, 4,300 had already died from the virus, more than double the number of deaths from the SARS and MERS viruses over the last fifteen years.

On March 13, 2020, President Trump declared a national emergency promising to provide significant disaster relief funding to state and local governments. The market soared during his speech, closing up 2,000 points. NBC NEWS

But many believed WHO was overreacting. For weeks, President Trump had tended to minimize the risks of the disease.

On February 10: "You know, a lot of people think that goes away in April with the heat—as the heat comes in. Typically, that will go away in April."

On February 25: "[China has] had a rough patch, and I think right now they have it—it looks like they're getting it under control more and more. They're getting it more and more under control. So I think that's a problem that's going to go away."

And on February 27: "It's going to disappear. One day—it's like a miracle—it will disappear. And from our shores, we—you know, it could get worse before it gets better. It could maybe go away. We'll see what happens. Nobody really knows."

Then the stock market collapsed, with the Dow Jones Industrial Average falling from 29,440 on February 14 to 19,028 on March 23. Radio and cable news pundits, feeding on the nation's political division, amplified the confusion, claiming COVID was a political hoax engineered by Democrats to harm the president during an election year.

On February 27, talk radio's most popular host, Rush Limbaugh, told his loyal listeners, "The forces arrayed against Donald Trump are doing everything they can to weaponize this to harm the economy, to harm the stock market in hopes of harming President Trump," Limbaugh declared. "Now, I want to tell you the truth about the coronavirus.... I'm dead right on this. The coronavirus is the common cold, folks."

Fox News soon picked up the political hoax drumbeat. On March 6, for example, Dr. Mark Siegel, the senior medical analyst for Fox News, told Fox viewers, "And let me tell you something, this virus should be compared to the flu, because at worst, at worst, worst case scenario it could be the flu." That same day, Sean Hannity amplified Siegel's comments telling his viewers "They're scaring the living hell out of people and I see it again as like, 'Oh, let's bludgeon Trump with this new hoax.'"

Not every pundit was declaring COVID-19 a hoax; some were cashing in. Within weeks of the outbreak, a mini-industry had popped up selling fraudulent COVID cures. On March 6, the Federal Trade Commission (FTC) sent warning letters to seven companies, which, the FTC claimed, were "preying on consumers by promoting products with fraudulent prevention and treatment claims." The companies were selling teas, essential oils, and colloidal silver, which they claimed would treat or prevent COVID-19. The FTC letter instructed the recipients to cease making claims that their products could treat or cure COVID-19.

Similarly, on March 12, the New York Attorney General sent a cease-and-desist letter to Alex Jones, the popular Infowars talk show host. The attorney general claimed Jones' Infowars and related websites promoted fraudulent COVID cures, including Superblue toothpaste, which the Infowars websites claimed "kills the whole SARS-corona family at point-blank range." The attorney general sent similar letters to other COVID cure promoters, including The Silver Edge company. In one of the more imaginative promotions at the time, Silver Edge used the former televangelist James Bakker to promote its Micro-Particle Colloidal Silver Generator along with Silver Wire for hundreds of dollars. The company claimed their silver products would not only cure but prevent COVID-19.

The New York Attorney General instructed Alex Jones to stop marketing Superblue toothpaste as a cure for COVID.

To his credit, Tucker Carlson, at the time Fox's second most popular host after Sean Hannity, was the voice of reason in the cacophony of false claims. Carlson believed his Fox associates and President Trump were dismissing the seriousness of COVID. On March 7, in an extraordinary two-hour meeting at Mar-a-Lago with President Trump, Carlson warned the president of the seriousness of the COVID-19 pandemic.

Two days later, Carlson told his audience, "People you know will get sick. Some may die. This is real. That's the point of this script—to tell you that." In a remarkable admission regarding President Trump's earlier COVID comments, Carlson continued, "People you trust—people you probably voted for—have spent weeks minimizing what is clearly a very serious problem. It's just partisan politics, they say, calm down. In the end this is just like the flu, and people die of that every year."

Carlson's comments were admirable and politically courageous, but the damage had already been done. By mid-

March, polling data showed that only 38 percent of Fox News viewers considered COVID a serious threat compared to 72 percent of national newspaper readers and 71 percent of CNN viewers. Overall, Pew Research found that 79 percent of people who turned to Fox News as their primary news source believed the media had exaggerated the risks of COVID.

On March 13, 2020—shortly after meeting with Carlson and two days after WHO declared COVID an international pandemic—President Trump declared a national emergency, giving his Secretary of Health and Human Services broad powers to manage the emerging pandemic. Although COVID had emerged less than three months earlier in Wuhan, China, nearly two thousand Americans had already been infected by the virus.

On March 13, 2020, President Trump declared a national emergency promising to provide significant disaster relief funding to state and local governments. The market soared during his speech closing up 2,000 points.

Three days later, President Trump issued stringent COVID mitigation guidelines, declaring, "My administration is recommending that all Americans, including the young and healthy, work to engage in schooling from home when possible. Avoid gathering in groups of more than ten people. Avoid discretionary travel. And avoid eating and drinking at bars, restaurants, and public food courts."

The president had been briefed on the spiraling death rates of earlier pandemics. Pandemics, if left unchecked, can quickly strain the nation's health facilities, flooding hospitals with the sick and dying.

In 1918, just six months after three flu-like deaths were reported in Haskell, Kansas, the Spanish Flu killed 195,000 Americans. Another 450,000 died over the next eighteen months. During the summer of 1957, the Asian Flu killed 116,000 Americans, and eleven years later, in 1968, nearly

100,000 Americans died from the Hong Kong Flu. Now with COVID infections growing exponentially, it was hoped that declaring a national lockdown would "flatten the curve," delaying the spread of the virus and giving hospitals and other facilities more time to prepare.

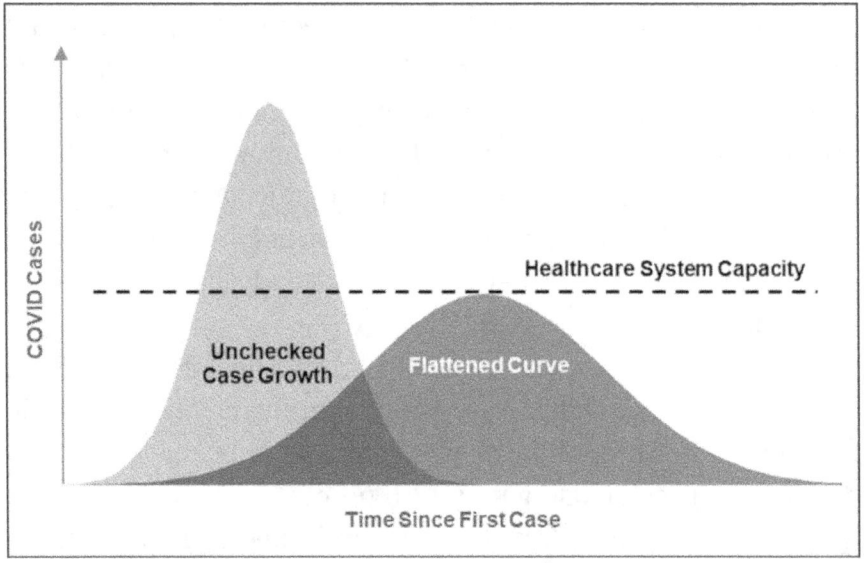

A rapidly spreading pandemic can easily overwhelm healthcare facilities and personnel. Slowing the disease's spread by restricting human interaction gives health officials time to prepare. CDC

A rapidly spreading pandemic can easily overwhelm healthcare facilities and personnel. Slowing the disease's spread by restricting human interaction gives health officials time to prepare.

Within days of President Trump's lockdown directive, communities began closing down, issuing restrictions on public gatherings, work, and travel. On March 15, New York City closed its public school system, the largest in the nation with 1.1 million students, followed the next day by closing bars, restaurants, gyms, and movie theaters. As the virus swept through the country, entire states began closing down, starting with California on March 19.

By April 7, forty-three states had largely locked down. Only seven states—Arkansas, Iowa, Nebraska, North Dakota, South Dakota, Utah, and Wyoming—opted not to issue lockdown orders.

The lockdowns may have slowed the virus, but they hardly stopped it. By early April, COVID was tearing through the country. On April 6, the death toll passed 10,000, an unimaginable number just a few weeks earlier. Five days later, deaths doubled to 20,000 on April 11, then doubled again on April 20.

As April ended, over 60,000 Americans had died of the virus, more than all the American deaths during the Vietnam War.

# Nine Millionaires and a Plumber

JANUARY 21, 2025

"Nine millionaires and a plumber" is how President Eisenhower's critics summed up his first cabinet. Eisenhower had never held elected office, or belonged to a political party, before his presidency. Consequently, with one exception, Eisenhower staffed his original cabinet with successful, and wealthy, businessmen rather than politicians. The exception was the Secretary of Labor, Martin Durkin, formerly president of the Plumbers and Steamfitter's Union.

These men were proud of their success, and contribution to America. Charlie Wilson, Eisenhower's Secretary of Defense and former CEO of General Motors spoke for many of his wealthy colleagues when he declared, "What's good for General Motors is good for America."

President Eisenhower's extended cabinet in 1955 was composed of some of the nation's most successful, and wealthiest, businessmen — and one woman.

Eisenhower's presidency has another link with President Trump. When Trump promotes "Make America Great Again" to what era is Trump actually referring? Certainly not any Democratic presidency. Nor either presidency of the two Bushes; young Bush gave America two Middle East wars topped off with the 2008 Financial Crisis while elder Bush was booted from office after breaking his "Read my lips, no new taxes" promise. Nor the Reagan presidency whose 1980 campaign slogan was "Let's Make America Great Again." Reagan, like Trump, was looking back to an earlier, golden era. (Trump appropriated Reagan's campaign slogan after dropping the collegial "Let's") And it certainly wasn't the Nixon era besmirched by the Watergate scandal.

During the nineteen-fifties, millions of Americans feared that communists had infiltrated every part of American society.     RKO FILMS

So, we are left with President Eisenhower and the nineteen-fifties; the earliest decade Trump and his generation can recall. For today's older Americans, the hazy fifties conjures a vision of peace and prosperity. The United States had won the Second World War in the process becoming the leader of the Free World. At home, traditional American families laughed at *I Love Lucy*, listened to Dinah Shore sing *See the USA in your Chevrolet*, shimmied with hula-hoops, and danced to rock-n-roll.

The fifties, then, must be the golden age to which Trump wishes to restore America.

Like Trump, who has inherited wars in Ukraine and the Middle East, Eisenhower inherited the Korean War, a war in which American troops were dying every day. Two months after winning the presidency in 1952, Eisenhower made a secret trip to Korea. Eisenhower, the top European commander during the Second World War, concluded that the war was unwinnable without a major escalation that would require, perhaps, the use of nuclear weapons to counter the massive Chinese army that was supporting their North Korean allies.

Eisenhower ended the fighting but was unable to end the war. A peace treaty was never signed. Instead, an armistice, little more than a temporary ceasefire, was signed on July 27, 1953. Legally, Korea has been frozen in a state of war for over seventy years.

Eisenhower, like Trump, also inherited a massive illegal immigration problem. In 1954, Eisenhower instituted "Operation Wetback" ("Wetback" was a common term for Mexicans who entered the United States illegally—often by swimming across the Rio Grande River). The Mexicans were largely workers who had been invited into the United States as part of the Bracero (Spanish for "laborer") program established during World War II. Mexico, not wishing to enter the war, provided the United States temporary laborers rather than military support. After the war, Washington turned a blind eye to the illegal workers who failed to return to Mexico. Like today, American farmers had become dependent on the hard-working, and low-wage, workers from south of the border.

Operation Wetback used military-style tactics to round-up and deport 1.3 million illegal Mexicans. Many soon returned.

But unlike today, by the mid nineteen-fifties, Mexico was suffering from its own labor shortages and demanded the laborers be returned. Eisenhower complied but declined to use

Federal troops to expedite the deportation. Instead, Eisenhower authorized border patrol agents to use military-style tactics to round-up and deport 1.3 million illegal Mexicans (proportionately equivalent to about 2.7 million today). The results were mixed; many of the deported Mexicans quickly returned to the United States.

More successful was President Eisenhower's efforts to build an Interstate Highway System.

It wasn't easy. Eisenhower proposed spending $50 billion ($590 billion in 2024 dollars) to construct the Interstate Highway System. The cost was massive. The total federal budget that year was only $71 billion.

Funding wasn't the only issue. Eisenhower's right-wing opponents claimed his national highway system was "creeping socialism," a denunciation heavily used during the Cold War. A few politicians even argued against free polio vaccines available to children as a "back-door" to socialized medicine. (Polio killed 3,145 children in 1955, crippled many thousands more, and condemned hundreds to an "iron lung").

But Eisenhower outsmarted his critics and their fear of a socialist takeover with a simple name change. The highway system would be renamed the National System of Interstate and *Defense* Highways. The new highways, Eisenhower asserted, would not only provide civilian transportation but would also be essential to evacuate cities and move military convoys during an atomic war.

That left the issue of how to pay for the massive new infrastructure. Unlike today, Eisenhower insisted that the highway program be "pay-as-you-go" and not contribute to budget deficits. This meant increasing fuel, tire, and other vehicle-related taxes—tax increases the automobile, oil, and trucking industries adamantly opposed.

The highway lobby had, for years, petitioned Congress to eliminate vehicle-related taxes and fund highway construction from general taxation sources, claiming that everybody, not just drivers, benefited from better roads. A gasoline tax increase was unthinkable.

For 35 years after the end of World War II, American presidents reduced the national debt as a percentage of GDP.

Ultimately, a compromise was reached to increase federal gasoline taxes from two cents to three cents a gallon ($0.35 in 2024 dollars). These taxes would be deposited into a Highway Trust Fund to assure they were spent on highways and not diverted for other purposes.

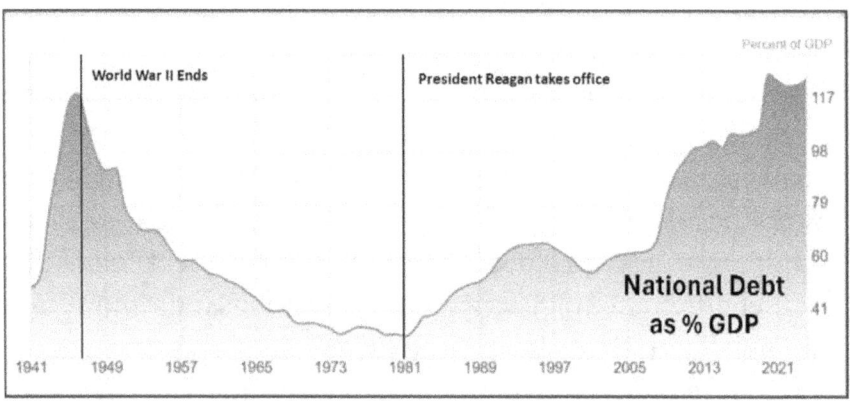

For 35 years after the end of World War II, American presidents reduced the national debt as a percentage of GDP.  FEDERAL RESERVE

Once the tax issue was resolved, the legislation flew through Congress. Six weeks later, on August 13, 1956, work began on the first stretch of road near St. Louis, Missouri. The final segment of the original highway plan wasn't completed until 1992 when I-70 opened through Glenwood Canyon, Colorado.

The Interstate Highway System revolutionized American travel and transportation. Contrary to the highway lobby's grim predictions, the automobile, trucking, and oil industries thrived after the Highway Act was passed. Rather than the trucking industry being driven out of business, the new superhighways fueled trucking's explosive growth: fifteen-fold from 1956 through 2006.

Concern over creeping socialism was a major political issue during the nineteen-fifties. Organizations such as the Keep America Committee and the John Birch Society claimed the fluoridation of public water supplies and the new polio vaccines were communist plots designed to slowly wipe out entire populations. The Indiana Textbook Commission even banned the book *Robin Hood* as communist. Robin Hood, as every schoolchild knows, robbed from the rich and gave to the poor, a concept the Indiana commission considered subversive. Book banning, though, wasn't just political. For decades, Boston had banned scores of books the city considered obscene including Walt Whitman's *Leaves of Grass* and Ernest Hemingway's *A Farewell to Arms*.

During the nineteen-fifties, millions of Americans feared that communists had infiltrated every part of American society.

The hunt for communist sympathizers reached its pinnacle under Senator Joseph McCarthy who brought hundreds of Americans before his House Committee on Un-American Activities (HUAC), often with little justification. McCarthy relied on degradation, intimidation, and the threat of imprisonment to extract testimony. Just being brought before the committee resulted in guilt by association, destroying reputations and careers. In Hollywood, an estimated three-hundred executives and artists were blacklisted. But others chose to cooperate, "not to save their lives, but to save their swimming pools," as film director Orson Welles dryly commented.

Eisenhower abhorred McCarthy's methods but chose not to challenge him directly—McCarthy enjoyed considerable support with many Americans. Instead, in April 1954, Eisenhower, in a brilliant political gambit, arranged to have McCarthy's investigation of the U.S. Army aired on live television.

Eisenhower knew his man. Millions of Americans soon saw McCarthy badgering and belittling American war heroes. When Lt. Colonel Chester T. Brown refused to answer a question, McCarthy fumed, "Any man in the uniform of his country who refused to give information to a committee of the Senate which represents the American people, that man is not fit to wear the uniform of his country."

McCarthy had finally gone too far. It was one thing to belittle a Hollywood screenwriter, but quite another to bully a decorated war hero. Joseph Welsh, the Army's legal counsel during the hearings, admonished McCarthy, "I think I never really gauged your cruelty or your recklessness. Have you no sense of decency, sir, at long last?" America had seen enough; McCarthy was little more than a common bully. The Senate agreed, and, in December 1954 censured McCarthy for his unsubstantiated accusations and violation of Senate behavioral standards (67 senators voted in favor of censure, 22 against). After the censure, McCarthy fell into obscurity. He died of liver failure in 1957.

During McCarthy's five years as chair of the Senate HUAC committee, not one person was sent to prison as a communist agent. By 1960, Americans were ready for a change and rejected Eisenhower's vice president, Richard Nixon, choosing instead the young John F. Kennedy as the nation's thirty-fifth president.

Historians didn't initially think particularly well of Eisenhower. For years they considered Eisenhower an affable but bumbling caretaker who spent much of his presidency golfing and fishing, an impression reinforced by Eisenhower's vibrant successor, John Kennedy, with his young family and eloquent, inspiring speeches.

Eisenhower's greatest quality was his quiet self-assurance. In 1982, after gaining access to Eisenhower's private presidential papers, Fred Greenstein wrote *The Hidden Hand Presidency*. The book described how Eisenhower intentionally avoided the spotlight, steering policy quietly through his staff. Eisenhower's definition of leadership was simple:

> "Leadership consists of nothing but taking responsibility for everything that goes wrong and giving your subordinates credit for everything that goes well."

# President Trump and Dr. Fauci Part Ways

JANUARY 28, 2025

By late April 2020, over 2,000 Americans were dying a day from COVID. The national lockdowns closing entire communities may have slowed the spread of the disease, but they had a high cost; lockdowns were devastating the economy. During February, even before President Trump's lockdown order, national employment fell by 1.4 million workers, the largest monthly decline since the end of the Second World War. The next month, the nation shed 20.5 million jobs throwing millions, largely hourly wage earners, out of work.

The lockdowns quickly generated an angry backlash. Prominent podcasters, radio and cable pundits, and even local public health officials began calling for their termination. During a "You Can't Close America" rally held in Austin, Texas on April 18, Infowars host Alex Jones called for an end to the lockdowns, declaring that "Texas is leading the war against the tyrants" while the crowd chanted "Fire Fauci," a reference to Dr. Anthony Fauci, Washington's most visible public health official.

Their anger was understandable. Texas had lost over one million jobs since the lockdowns began. But the beleaguered Fauci had no control over lockdowns in Texas or any state; those were ordered by state and local officials who quietly let Fauci take the blame. Indeed, seven states — Arkansas, Iowa, Nebraska, North Dakota, South Dakota, Utah, and Wyoming —

never locked down and chose to remain open throughout the pandemic.

Potentially more dangerous were protests held in Lansing, Michigan days later when heavily armed protestors entered the Michigan statehouse during an "American Patriot Rally." Several of the protestors carried signs paraphrasing Benjamin Franklin's oft-quoted dictum, "Those who would give up essential Liberty, to purchase a little temporary Safety, deserve neither Liberty nor Safety."

A photo taken during a lockdown protest at the Arizona state capitol captured the tension between personal freedom and public health.

IMAGN

For millions of Americans, Franklin's phrase captured their resistance to government mandates. But Benjamin Franklin's famous quote is commonly misunderstood. The phrase concerned a tax dispute, not personal freedom. Franklin included the phrase in a 1755 letter he had authored on behalf of the Pennsylvania Assembly to the colonial governor.

At the time, the French and Indian War was raging threatening Pennsylvania settlers on the colony's western border. In the letter, Franklin urged the governor not to accept a single lump sum payment—essentially a bribe—offered by the Penn family who had founded the colony. The money was to be used to fortify the western border (temporarily purchasing "safety"). In exchange, the Penn family demanded the governor permanently forfeit the colony's authority to tax the family (the loss of "liberty").

Ironically, Franklin's famous phrase was in support of continued taxation, not personal liberty. Still, over two centuries later, Franklin's phrase perfectly captured the tension between public health and personal freedom that divided the nation during the pandemic.

Unlike earlier pandemics, COVID was killing the elderly at far higher rates than those younger. During the first three months of the pandemic, over 80 percent of COVID deaths were sixty-five years or older, an age group which constituted only 17 percent of the population. Yet nearly the entire nation was locked down.

An early advocate for ending lockdowns and restoring to millions of Americans their personal freedoms was Dr. Scott W. Atlas, a senior fellow at Stanford University's Hoover Institution and the former chief of neuroradiology at Stanford University Medical Center. In an April 22, 2020, opinion piece published in *The Hill*, Atlas argued for a more focused approach to managing the COVID-19 pandemic. Recognizing that COVID primarily killed the elderly and those with underlying conditions, Atlas proposed that these vulnerable groups be isolated from the general population, allowing schools, businesses, and other institutions to remain open.

But during the early months of the pandemic, public health officials remained cautious. They knew earlier influenza pandemics had killed the young and elderly at similar rates; the elderly due to weakened immune systems and frequent comorbidities such as heart and respiratory disease, while the young, with undeveloped immune systems, lacked natural immunity.

In 2009, a Swine Flu pandemic killed over 12,000 Americans, 80 percent of whom were under 65 years of age. And as every epidemiologist knew, in less than two years the 1918 Spanish Flu killed 650,000 Americans; the large majority of whom were under forty years old and in the prime of their lives.

If the COVID virus wasn't currently killing the young, virologists knew a mutation could quickly change that.

By June, after months of restrictions and with summer approaching, Americans were anxious to resume their normal lives. COVID death rates had declined to an average of 770 a day, an unimaginable rate just a few months earlier, but well below their April average of nearly 2,000.

The CDC continued to urge caution as new COVID hot spots popped up around the country. "States in the South, West and Southwest," *The New York Times* reported on June 15, "are seeing upticks in their COVID case counts—and in some cases setting records—as a troubling pattern emerges in areas that began lifting restrictions earlier than others."

Was *The New York Times* over-reacting? From the beginning of the pandemic, skeptical cable and social media pundits claimed the COVID deaths were inflated, or not even COVID at all, just higher than normal influenza or other deaths.

But that was a difficult argument to support. During the ten years from 2010 through 2019, U.S. deaths grew steadily at an average rate of 1.6 percent per year largely due to the nation's aging population. But in 2020, the first year of the pandemic, U.S. deaths increased an astounding 18.7 percent resulting in 464,000 more deaths than projected. Statisticians classify the increase in actual deaths over projected deaths rather coldly as excess deaths.

Many types of deaths increased in 2020, from heart attacks to traffic accidents for a variety of reasons. COVID often killed those with multiple comorbidities making it difficult at times to identify the specific cause of death. "If somebody has end stage pancreatic cancer and COVID, did they die with COVID or from COVID?" John Fudenberg, a medical examiner, asked rhetorically in a November 2020 medical journal. Certainly, some deaths were misclassified. Still, the overwhelming majority of excess deaths, by all credible accounts ranging from the nation's top hospitals to small town coroners, were due to COVID.

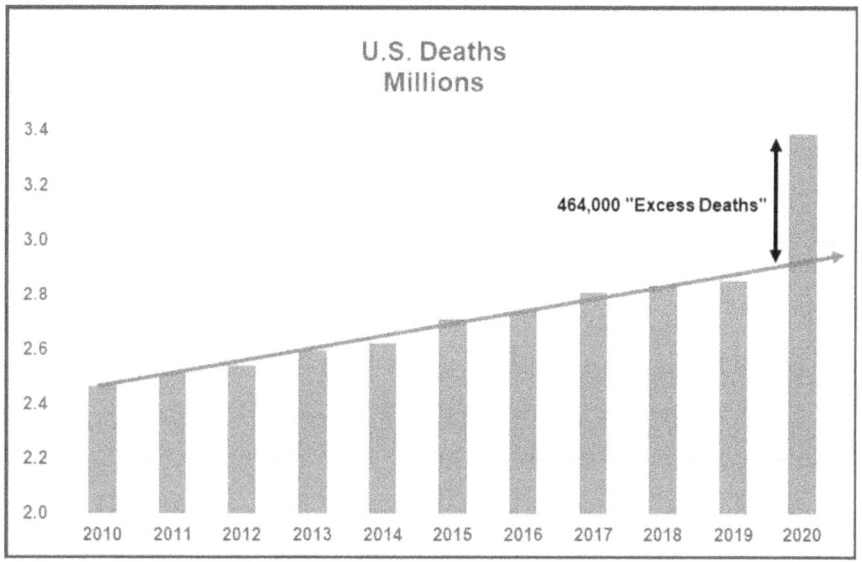

The increase in U.S. deaths during 2020 was the largest since the 1918 Spanish Flu pandemic. CDC

The day after *The New York Times* article warned of a resurgence of COVID deaths, Vice President Pence responded with a strong rebuttal in *The Wall Street Journal* entitled, "There Isn't a Coronavirus Second Wave." "In recent days," Pence wrote, "the media has taken to sounding the alarm bells over a 'second wave' of COVID infections. Such panic is overblown. The media has tried to scare the American people every step of the way, and these grim predictions of a second wave are no different. The truth is, whatever the media says, our whole-of-America approach has been a success."

President Trump agreed. Encouraged by the decline in deaths, on June 25, the president tweeted, "Coronavirus deaths are way down. Mortality rate is one of the lowest in the World. Our Economy is roaring back and will NOT be shut down. 'Embers' or flare ups will be put out, as necessary!"

President Trump's concern for COVID's economic impact was well-founded but his claim that America's mortality rate was one of the lowest in the world was widely off the mark. Among the developed nations, the U.S. COVID death rate was second only to Sweden's towering death rate. During June 2020, America's monthly death rate was 58 COVID deaths per million, below Sweden's 89 deaths per million, for sure, but far above other developed nations, including Spain (26 deaths per million), Canada (23), France (15), Germany (6), and Switzerland (1.5).

Dr. Anthony Fauci strongly objected to the president's claim that the pandemic was receding. On June 30, Fauci, testifying before a Senate panel, voiced his concerns:

> "Clearly, we are not in total control right now.... It is going to be very disturbing, I will guarantee you that, because when you have an outbreak in one part of the country, even though in other parts of the country they're doing well, they are vulnerable. We are now having forty-plus thousand new cases a day. I would not be surprised if we go up to 100,000 a day if this does not turn around."

Dr. Fauci was unfortunately right. A new, more deadly COVID strain, Alpha, had begun to emerge over the summer. After declining in June, 60,000 new COVID cases were reported daily in July rising steadily to 100,000 in November then quickly doubling to 200,000 in December overwhelming hospitals and their medical staff.

# Alternative Facts

FEBRUARY 4, 2025

Charlatans and conspiracists thrive during periods of uncertainty and fear, and few events breed uncertainty and fear more than a pandemic that kills indiscriminately. That was true in 1665 during London's Great Plague, when "the posts of houses and corners of streets were plastered over with doctors' bills and papers of ignorant fellows, quacking and tampering in physic, and inviting the people to come to them for remedies."

It was also true during the first modern influenza pandemic, which began in the Russian Empire in 1889. A year after emerging from Russia, the virus had killed a million people, equivalent to over five million deaths today.

Although medical science was rapidly advancing in the late nineteenth century, the causes of most diseases, including influenza, were still unknown. Around 1880, germ theory—that disease was caused by tiny microbes—had begun to emerge. Germs were easily seen under a microscope, but where did they come from?

An improbable theory soon emerged: that electricity, the astonishing new technology being harnessed by Thomas Edison and Nikola Tesla, somehow gave birth to microbes. On January 31, 1890, the Paris edition of *The New York Herald* wrote, "The invention of electric light has been followed by the appearance of a microbe that employs its spare time in producing Russian influenza. The disease has raged chiefly in towns where the electric light is in common use and has penetrated slowly and reluctantly into towns where the electric lamps are unknown." The article noted that the disease "has everywhere attacked telegraph employees."

It would be decades before researchers understood that the pathogens that cause disease have existed for millennia, evolving and mutating, passing from species to species, usually harmlessly, but occasionally killing their host.

The Carbolic Smoke Ball was a popular cure for many afflictions during the 1892 Russian Plague.   WELLCOME COLLECTION

Like the London Plague more than two-hundred years earlier, charlatans quickly arose during the 1892 Russian Plague selling quack remedies. The Carbolic Smoke Ball, a rubber ball filled with powdered carbolic acid, was the most notorious, promising to prevent influenza and myriad other diseases. Users were instructed to squeeze the ball, sending a puff of carbolic acid through a small rubber tube into their nostrils. The product was endorsed by famous personalities and backed by a £100 guarantee.

When a customer, after using the product diligently for three months, contracted influenza, the company refused to pay the guarantee, then lost the lawsuit after the deceived customer sued. The case set a precedent for truth in advertising, declaring a published offer is a legally binding contract. Today, related laws require that the claims promoters make regarding medical treatments and cures must be truthful.

More than a century later, as COVID was beginning its assault on the world's population, any individual with access to a microphone or keyboard could freely promote their often reckless views to a worldwide audience on the pandemic, vaccines, or the latest global conspiracy.

Theories quickly sprouted to explain the origins of the COVID-19 pandemic.

On May 4, 2020, Mikki Willis, an independent film producer, released *Plandemic: The Hidden Agenda Behind Covid-19* on Facebook and other social media. The *Plandemic* video claimed that global elites, including the World Health Organization and Microsoft founder Bill Gates, had engineered the pandemic to profit from selling vaccines. The video was financed by Children's Health Defense, a nonprofit organization headed by Robert F. Kennedy Jr.

A second theory regarding the pandemic's origin was promoted by, of all people, Archbishop Carlo Maria Vigano. The outspoken archbishop had spent much of his career investigating financial and sexual scandals within the Vatican. In October 2020, Archbishop Vigano published an open letter to President Trump which alleged a gigantic global conspiracy:

> "A global plan called the Great Reset is underway. Its architect is a global élite that wants to subdue all of

humanity, imposing coercive measures with which to drastically limit individual freedoms and those of entire populations.... Behind the world leaders who are the accomplices and executors of this infernal project, there are unscrupulous characters who finance the World Economic Forum and Event 201, promoting their agenda."

Event 201 was a simulation of a worldwide pandemic conducted in 2019 by the Johns Hopkins Center for Health Security in partnership with the Bill and Melinda Gates Foundation and World Economic Forum. The outbreak of the COVID-19 pandemic just months after the Event 201 simulation spawned claims that the pandemic was a planned event, that an evil global cabal had orchestrated the worst public health crisis in a century.

The claim that the pandemic was a planned event had strong supporters, including Robert F. Kennedy Jr. During a May 2022 podcast interview, Kennedy claimed, "A global elite led by the CIA had been planning for years to use a pandemic to end democracy and impose totalitarian control on the entire world."

The Plandemic and Great Reset were fantastic global conspiracies. Other conspiracy claims were less grandiose but equally damaging. Rather than trusting long respected health institutions or even their personal physician, many chose to rely on self-appointed, social media pundits.

On July 27, 2020, America's Frontline Doctors (AFD) held a press conference on the steps of the Supreme Court. AFD was founded by Dr. Simone Gold, a licensed emergency-room physician and Stanford-educated lawyer.

Gold had founded AFD to advocate alternative approaches to treating COVID. During their "White Coat Summit," Dr. Gold and her associates promoted the anti-malaria drug hydroxychloroquine as a treatment for COVID. Many Americans had become inclined to try the unproven drug after

President Trump promoted it during a White House event in May. But on June 15, the U.S. Food and Drug Administration revoked its emergency authorization for hydroxychloroquine as a treatment for COVID. Studies had shown the drug was ineffective treating the disease and, in rare instances, led to fatal heart arrhythmias.

Dr. Gold and her white-coated associates staged the press conference to announce that they disagreed with the FDA. Although the press conference was sparsely attended, a video of the event went viral after President Trump retweeted it, along with other tweets that day claiming, "Hydroxychloroquine could save up to 100,000 lives if used for COVID-19."

Interest in hydroxychloroquine, though, was soon replaced by ivermectin, a drug used to treat parasitic worms. America's Frontline Doctors quickly pivoted to support ivermectin as a cure for COVID, claiming the COVID vaccines were "experimental biological agents" and "not effective in treating or preventing COVID-19."

Although America's Frontline Doctors and other ivermectin marketers made millions promoting alternatives to COVID vaccines, ivermectin never received mainstream approval. On February 4, 2021, ivermectin's manufacturer, Merck, issued a statement declaring that there was "no scientific basis for a potential therapeutic effect against COVID-19 from preclinical studies [and] no meaningful evidence for clinical activity or clinical efficacy in patients with COVID-19 disease."

Even more damning was India's decision to drop both hydroxychloroquine and ivermectin as COVID treatments. India, a nation of 1.4 billion people, was desperate to find an inexpensive, effective treatment for COVID and had hoped these drugs were the answer.

They weren't. On September 26, 2021, the India Today website reported that the [Indian] National Task Force on Covid-19, after months of clinical trials, had "dropped the use of Ivermectin and Hydroxychloroquine drugs from their revised guidelines for the treatment of the [COVID] infection.... The decision was taken after experts found that these drugs had little to no effect on Covid-related mortality or clinical recovery of the patient."

From the beginning of the pandemic, pundits, quacks, and conspiracists convinced millions of Americans that COVID was a political hoax or no worse than the common flu; that the vaccines were more dangerous than the disease itself; that ivermectin and hydroxychloroquine were COVID cures; and that the pandemic was an evil, worldwide plot masterminded by global elites. Five years, and two Congressional investigations, after the beginning of the pandemic, no credible evidence has surfaced that any of these sensational claims are true.

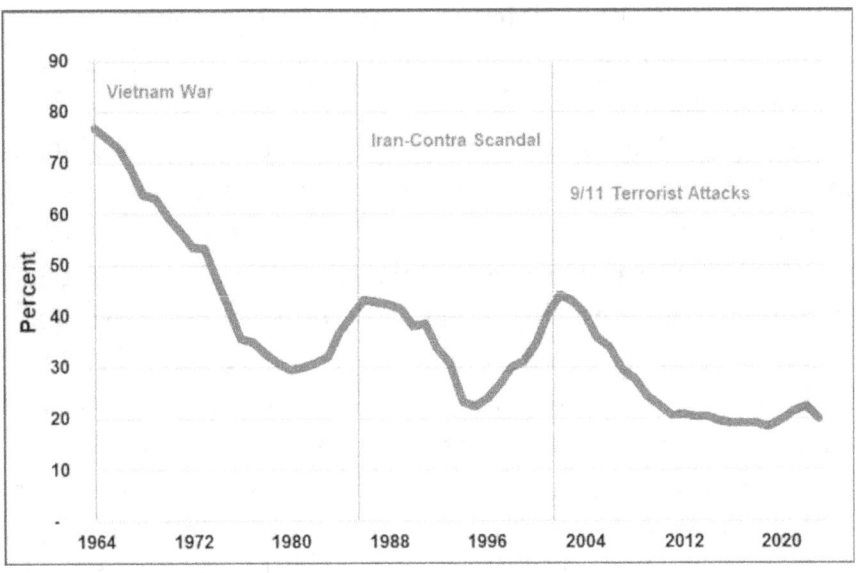

By 2020, only 20 percent of Americans trusted their government to do the right thing "all or most of the time." GALLUP

"Everyone is entitled to his own opinion, but not his own facts," as social commentator and senator Daniel Patrick Moynihan famously declared. For generations, journalists took professional pride in being factual. Hard-nosed newspaper editors told their cub reporters, "If your mother says she loves you, go check it out."

But many of today's information sources are awash in "alternative facts," that memorable phrase coined by Kellyanne Conway, senior counselor to President Trump. Conway was struggling to defend President Trump's false claim that his 2017 inauguration crowd was the largest in history. With no supporting facts, Conway simply presented Trump's grandiose claims as alternative facts.

But alternative facts are largely the result, not the cause, of today's disinformation. The root issue is Americans' distrust in their nation's institutions. Distrust in traditional sources—the government, media, and the nation's medical, scientific, and academic institutions—has caused millions of people to become susceptible to alternative information sources, especially sources that feed their existing biases and suspicions.

# Dr. Atlas and America's Giant Measles Party

FEBRUARY 11, 2025

On June 30, 2020, Americans were asked to choose between President Trump and Dr. Anthony Fauci, his top public health official. Five days earlier, the president had declared "Coronavirus deaths are way down. Mortality rate is one of the lowest in the World. Our Economy is roaring back...." After three months of lockdowns, President Trump was anxious to reopen America.

But Dr. Fauci, disagreed. Worse, he disagreed publicly. On June 30, Fauci testified before a Senate panel. "Clearly we are not in total control right now..." Fauci told the senators. "We are now having forty-plus thousand new cases a day. I would not be surprised if we go up to 100,000 a day if this does not turn around."

Fauci had not only publicly contradicted the president, Fauci's prediction had been right. By the end of the year, 345,000 Americans would be dead from COVID, a large majority having died during the six months after Fauci made his prediction.

For months, President Trump had been uncomfortable with several of his COVID Task Force members. That was especially true for the outspoken Dr. Anthony Fauci, who was disparaged as "Dr. Doom and Gloom" by many of the president's aides inside the White House.

During July, the White House began a campaign questioning Dr. Fauci's management of the pandemic. In mid-July, the White House quietly distributed a paper to friendly media outlets smearing Dr. Fauci. Similar in spirit to President Nixon's "Dirty Tricks" decades earlier, the document provided examples of Fauci purportedly offering bad medical advice. The accusations, though, were often taken out of context or relied on selective editing.

A cartoon posted by the White House staff was part of a smear campaign against Dr. Fauci.     WHITE HOUSE

That same month, the White House quietly began a social media campaign against Dr. Fauci starting with a cartoon rendering Fauci as a faucet drowning Uncle Sam. The cartoon was posted by Daniel Scavino, a White House deputy chief of staff.

Watching from a distance, Canada's *Toronto Star* newspaper declared that "the smearing of Fauci is beyond disgusting."

Dr. Fauci's popularity with the public may have contributed to his unpopularity in the White House. From the beginning of the pandemic, public opinion polls consistently rated Fauci's handling of the COVID-19 pandemic well above President Trump's. A Quinnipiac University survey conducted in May, at the height of the first COVID wave, asked respondents whether they approve of President Trump's and Dr. Fauci's handling of the coronavirus. The respondents gave Dr. Fauci a strong 68 percent approval rating while Trump was rated at 41 percent.

On August 10, President Trump introduced Dr. Scott Atlas, a respected Stanford professor and radiologist, as the newest member of his COVID Task Force. Starting with an April opinion piece published in *The Hill*, Dr. Atlas had been a frequent commentator on Fox News and other conservative outlets, advocating for eliminating most COVID restrictions. The favorable coverage caught the president's attention. "Scott is a very famous man, who is also highly respected," Trump said. "He's working with us and will be working with us on the coronavirus. And he has many great ideas."

Dr. Atlas believed the nation could achieve "herd immunity" to COVID by allowing the young and healthy to become infected while protecting those at most risk. Months before his White House appointment, Dr. Atlas had described his approach during an April 23 interview on the popular Steve Deace Show:

> "We can allow a lot of people to get infected, those who are not at risk to die or have a serious hospital-requiring illness, we should be fine with letting them get infected, generating immunity on their own, and the more immunity

in the community, the better we can eradicate the threat of the virus."

Dr. Atlas's approach to herd immunity reminded older Americans of the popular "measles parties" held during the nineteen-fifties. Before reliable vaccines were available, parents intentionally exposed their children to measles and other childhood diseases (while eating cake and ice cream) in the hope that the children, after a brief infection, would build-up lifelong immunities to the diseases. The parties were a calculated risk by parents, as during the nineteen-fifties, five hundred American children typically died of measles each year. Today, with safe and effective vaccines available, public health officials strongly discourage measles parties. Although few die of measles today, it remains a serious disease, with one in four infected individuals requiring hospitalization.

Dr. Atlas's proposal to encourage the virus to spread throughout the population was opposed by much of the medical community, including many of his former Stanford colleagues. On September 9, seventy-eight medical professionals at Stanford Medical School wrote an open letter condemning Atlas. The letter claimed Atlas was promoting "falsehoods and misrepresentations of science [which run] counter to established science and, by doing so, undermine public health authorities and the credible science that guides effective public health policy."

But millions of Americans, exhausted with COVID restrictions and anxious to return to their normal lives, agreed with Dr. Atlas. Perhaps none more so than the nation's hardcore motorcyclists. From August 7 through August 16 over 460,000 motorcyclists converged on Sturgis, South Dakota for the 80th Annual Sturgis Motorcycle Rally. Few were worried about social distancing and face masks. Why should they be after riding across the country on a motorcycle at seventy miles per hour?

The rally was filled with ten days of motorcycle races, parades, contests, vendor fairs (T-shirts proclaiming "Screw COVID, I Went to Sturgis" sold well.), rock concerts, and all-around carousing.

But more than a giant block party, for many motorcyclists the Sturgis Rally was a cherished annual ritual. Chuck Chamberlain had attended the rally since he was twenty-four years old. "My first year was 1986," Chamberlain told a *New York Times* reporter. "This is my family. We communicate for the rest of the year. My children come here, my brother comes here, my daughter met her husband here. My daughter got married here." For thousands like Chamberlain, the warm camaraderie of Sturgis outweighed the risks of COVID.

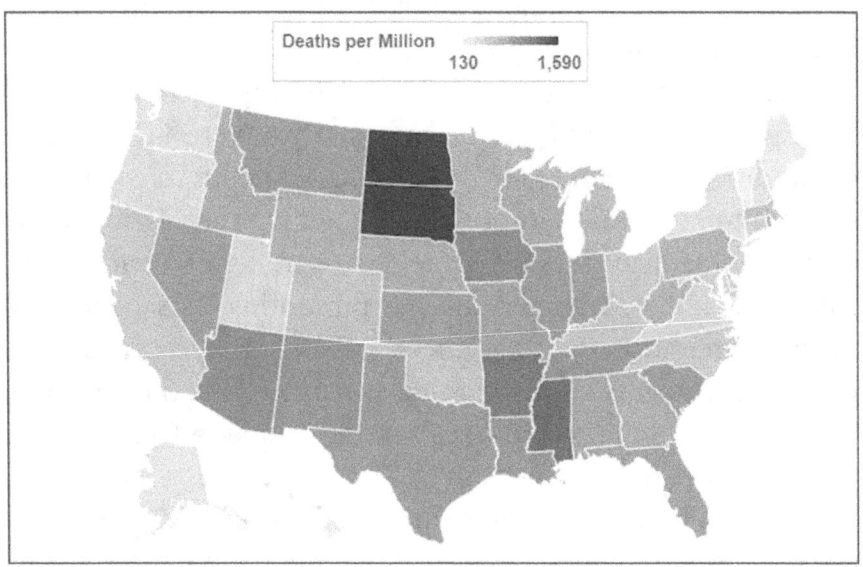

From July through December 2020, North and South Dakota had the nation's highest COVID death rates. CDC

The urge to return to a normal life was understandable, but epidemiologists feared large gatherings like the Sturgis Rally would increase COVID deaths, both among the rally attendees, and later friends and family infected when the motorcyclists returned home. A month after the rally, a study conducted by San Diego State University used cell phone data to track rally attendees returning to their home in counties across the nation. Based on the rise in COVID cases in those counties, the study estimated that 267,000 new COVID cases were linked to the rally.

Some researchers believed the study's estimate of 267,000 new COVID cases was far too high. But a month after the Sturgis Rally, COVID death rates in North and South Dakota began rising steeply. By the end of the year, the two Dakotas, measured during the period from July through December, had suffered the nation's highest COVID death rates. Together, North and South Dakota averaged over 1,500 deaths per million while the other forty-eight states averaged only 650 deaths per million.

There's no available record of what Dr. Atlas thought regarding the Sturgis Motorcycle Rally, but South Dakota Governor Kristi Noem stood firmly behind her support for the rally. In a late August radio interview, Governor Noem was asked whether the rally should have been allowed: "I think it was an event that a lot of people came and enjoyed and exercised their freedom," Governor Noem responded. "We love South Dakota and think it's beautiful and tourism is our number two industry."

Most South Dakota residents must have agreed. Even though the state's COVID death rates during 2020 were among the highest in the nation, Governor Noem won reelection two years later by a 27 percent margin.

# Harding, Trump, and America First

FEBRUARY 18, 2025

Harding's lucky break came one day in 1899 when he happened to need a shoeshine.

Sitting next to him on the shoeshine stand was Harry Daugherty. As Harding stepped down and cordially tipped the bootblack, Daugherty had a life-altering insight. Harding had the genuine affability, dignified bearing, and exceptionally good looks that could be molded into a president. At twenty-nine, Daugherty already knew he would never rise to high political office on his own merits; his real calling was as a kingmaker. On that day, Daugherty found his future king. For the next twenty years, Daugherty would help propel Harding's political career from smalltown newspaper publisher to president of the United States.

Likable and non-controversial, Harding progressed rapidly, beginning with his election to the Ohio Senate in 1900, then to lieutenant governor, and finally to the U.S. Senate. By 1920, Harding, with Harry Daugherty's shrewd help, was running a longshot bid for president. Harding campaigned on *America First*. His campaign slogan was well-timed. Americans were dispirited and exhausted. In the preceding three years, 117,000 American soldiers had died in the First World War, the Spanish Flu had killed 675,000 Americans, inflation had ballooned to

20 percent, and hundreds of thousands of foreign refugees had flooded into the country.

Harding attributed many of these problems to America's drift away from what he called *Americanism*. "Call it the selfishness of nationality if you will," Harding declared in a 1920 campaign speech, "I think it's an inspiration to patriotic devotion to safeguard America first..."

During the 1920 presidential campaign, Warren G. Harding ran on a campaign promise of America First.  WHITE HOUSE

After eight years of the cerebral President Woodrow Wilson (Wilson had formerly been the president of Princeton University), voters liked Harding's down-to-earth, masculine style. That was especially true for America's women who were voting for the first time after the passage of the 19th Amendment.

Harding won the 1920 election with 60 percent of the popular vote, crushing his opponent, James Cox, by twenty-six points—to this day, the largest popular vote margin ever amassed by a president since the beginning of today's two-party system in 1824.

Harding's presidency foreshadowed two of today's most divisive issues: immigration and foreign entanglements.

In May 1921, Harding signed the Emergency Quota Act to limit the influx of Eastern and Southern Europeans—namely Jews, Poles, and Italians—fleeing their homes after the First World War and the Russian Revolution.

Harding's 1921 Act was a precursor to the much more stringent National Origins Act of 1924 under President Coolidge. The purpose of the 1924 Act was to preserve what Congress considered to be the ideal racial homogeneity of the American population, namely Western and Northern European. All Asians, from China to India to Japan, were excluded from immigration. It wasn't until 1952 under the McCarran-Walter Act that Asians were allowed to become naturalized U.S. citizens. (Chinese, American allies during World War II, were granted naturalization rights in 1943.)

The need to maintain America's racial homogeneity was broadly accepted during the first half of the twentieth century. Yet, Harding was the first sitting president since Reconstruction to take a message of racial equality to the Deep South when he spoke to a segregated crowd of 100,000 in Birmingham, Alabama on October 21, 1921.

"I can say to you people of the South, both white and Black..." Harding declared as he pointed at the white section of the audience, "whether you like it or not, our democracy is a lie unless you stand for that equality.... I would say let the Black man vote when he is fit to vote; prohibit the white man voting when he is unfit to vote." Half the crowd cheered, the other sat in stony silence. One congressman denounced Harding's speech as "a blow to the white civilization of America."

Harding also urged Congress "to wipe the stain of barbaric lynching from the banners of a free and orderly, representative democracy." Congress, as it had for decades, failed too act. Few presidents other than Harding have had the courage to speak out strongly against lynching; the southern voting bloc was too strong. By 2021, nearly 200 anti-lynching bills had been submitted since the Civil War; none made it through both houses of Congress until President Biden signed the Emmett Till Antilynching Act in 2022.

Although President Harding was liberal in his outreach to America's Black citizens, he was conservative in his approach to international relations.

As the First World War was coming to an end, President Wilson presented his Fourteen Points plan to assure a peaceful transition after the war. One of Wilson's fourteen points proposed the establishment of the League of Nations, "a general association of nations... to mediate international disputes."

The League of Nations was officially founded in 1920 with forty-two charter members, but President Harding refused to join. Harding had gauged the temperament of the American people. After the devastation of World War I, Americans wanted to avoid involvement in European affairs; Harding's *America First* policy meant the nation would focus on domestic issues rather than international commitments.

Harding's *America First* and its attendant isolationism would remain popular for two decades until Japan's surprise attack on Pearl Harbor on December 7, 1941. Overnight, the United States transitioned from isolationism to the global leader in international affairs, a role America would maintain for the next seventy-five years.

Perhaps Harding's greatest contribution as president was his appointment of two exceptional men to key positions in his administration: Andrew Mellon as Treasury Secretary, and Herbert Hoover as Commerce Secretary. Their economic and political impact would ripple through America for the next century.

Mellon argued that high tax rates caused the rich to avoid paying taxes by hiding income, employing tax shelters and even working less since much of their additional income would be taxed away. By lowering tax rates, Mellon claimed more taxes would be collected. He proposed cutting the top income tax rate from 73 to 25 percent. "It seems difficult for some to understand that high rates of taxation," Mellon later wrote, "do not necessarily mean large revenue to the Government, and that more revenue may often be obtained by lower rates." Fifty years before the American economist, Arthur Laffer, expropriated it, Mellon had defined the "Laffer Curve," a key element of today's conservative tax policies.

Mellon, one of America's most respected capitalists, quickly sold Congress on his tax plan. Eight months after his appointment, Congress approved the Revenue Act of 1921. To encourage private investment, Congress slashed taxes on capital gains to 12.5 percent, their lowest rate for the next one hundred years. Mellon called his plan "scientific taxation." Today, we know Mellon's tax policies as supply-side economics.

Herbert Hoover served as Commerce Secretary under both Presidents Harding and Coolidge. Hoover was a hyper-active Commerce Secretary who drove initiatives in movies, radio, aviation, highway safety, product standardization, home ownership, and flood control. Hoover's accomplishments included the Air Mail Act of 1925, which was the beginning of the airline industry. The Radio Act of 1927 established the precursor to the Federal Communications Commission. During these years, Hoover's Commerce Department worked with industry to standardize tools, hardware, and packaging to make manufacturing and shipping more efficient.

But beyond commerce, Herbert Hoover's primary contribution to President Harding's Americanism was a short booklet he wrote in 1922. Titled *American Individualism*, Hoover's book argues that the success of the United States is rooted in its unique system of individual initiative, personal liberty, and self-reliance—what Hoover later described as America's rugged individualism.

Hoover's notion of rugged individualism pervades America's self-image, from John Wayne in the classic movie *"The Searchers"* to John Galt, the personification of rational self-interest in Ayn Rand's novel *Atlas Shrugged*, to today's billionaire entrepreneurs. It's an image America much of America proudly embraces.

Unfortunately, Harding also appointed friends and political cronies, dubbed the "Ohio Gang," to key positions: Harry Daugherty as Attorney General, Albert Fall as Interior Secretary, and Charles Forbes as Director of the Veterans' Bureau, along with a motley collection of drinking buddies, golf partners, and general sycophants.

Two years into his presidency, Harding learned that many of the Ohio Gang had betrayed his trust by using their government positions to enrich themselves. "I have no trouble with my enemies," Harding confided to a newspaper editor, "But my damn friends, my God-damned friends… they're the ones who keep me walking the floor nights!" Seeking respite from Washington, Harding began an extended speaking tour throughout the American West and Alaska.

Harding never returned to Washington. Depleted by worry and weakened earlier by a virulent case of the flu, Harding died of a heart attack on August 2, 1923, while in San Francisco.

No President since Lincoln was more deeply mourned. An estimated nine million Americans came out to honor his funeral train as it made its way from California back to Washington. As the train passed, mourners sang Harding's favorite hymn, "Nearer My God to Thee." America's 19,134 newspaper boys contributed one copper penny each to be melted down and sculpted into a statue of Laddie Boy, Harding's beloved dog.

After his death, Harding's presidency was scandalized as evidence of broad political corruption within his administration—but not a hint by Harding himself—emerged. But the greatest damage to Harding's reputation came not from the Ohio Gang, but from a young woman. In 1927, Nan Britton published *The President's Daughter*, which claimed Harding had fathered her daughter, Elizabeth Ann, in 1919 during a tryst in Harding's Senate office. Harding's family disputed the claim, but in 2015, a DNA test on Britton's grandson confirmed that Elizabeth Ann Britton was, indeed, the daughter of Warren G. Harding.

Warren and Florence Harding with their beloved dog, "Laddie Boy."
FPG | GETTY

Today, those scandals are largely forgotten. What remains are the political philosophies developed under President Harding: *America First*, limited immigration, and supply-side economics. Philosophies which have underpinned Donald Trump's two presidencies and his promise to *Make America Great Again*.

# Vaccines: a Mighty and Horrible Monster

FEBRUARY 25, 2025

A smallpox epidemic nearly cost America its freedom. Throughout the Colonial period, smallpox regularly swept through the Colonies. Boston alone suffered smallpox outbreaks in 1721, 1730, 1752, and 1764. As America's War for Independence erupted in 1775, smallpox was yet again emerging in Massachusetts.

In November 1775, a Continental Army marched northward from Boston to Quebec City with orders to wrest the city from the British. Traveling with the army was smallpox. By late December, most of the army's 3,200 soldiers were suffering from the disease and unfit for duty. The British weren't nearly as affected. Smallpox, the "speckled monster" as it was known in Britain, had ravaged Europe for centuries, providing Europeans, and British soldiers, a degree of natural immunity.

On December 30, the desperate Continental Army attacked the city. The Americans were quickly routed, and hundreds of Colonists captured. Their commander, General Richard Montgomery, died in the battle. Hoping for reinforcements, the surviving army remained in the region. Three months later, Major General John Thomas arrived to command the remaining troops. Within days, Thomas died from smallpox. Fresh British troops soon landed to reinforce the city, leaving the ruined Continental Army little choice but to flee.

Despondent, John Adams, the future president, wrote his wife, Abigail: "Our Misfortunes in Canada, are enough to melt a Heart of Stone. The Small Pox is ten times more terrible than Britons, Canadians, and Indians together. This was the Cause of our precipitate Retreat from Quebec…"

Thanks largely to smallpox, Canada would remain British.

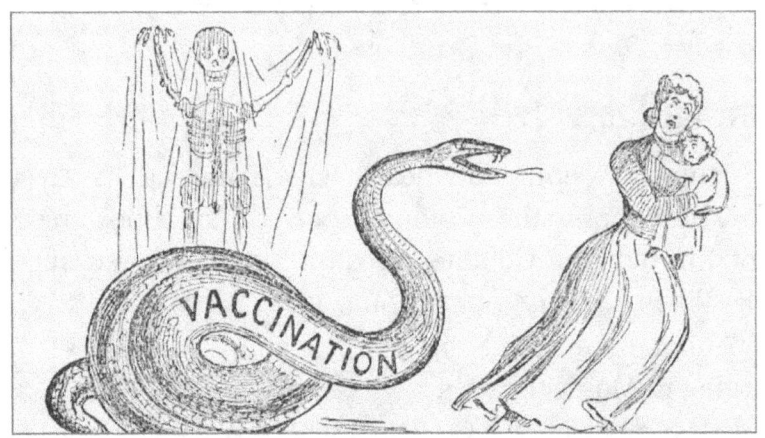

An anti-vaccination cartoon published in 1894.

By February 1777, General George Washington had concluded that only inoculation, an early form of vaccination, would prevent his army from being destroyed. Many states, though, had banned inoculation, primarily for religious reasons, forcing Washington to ask their governors for permission. Petitioning Virginia Governor Patrick Henry, Washington wrote that smallpox "is more destructive to an Army in the Natural way, than the Enemy's Sword."

Fortunately, the governors granted General Washington consent to inoculate the army. Had they not, Washington's Colonial Army may well have suffered the same fate as General Montgomery's in Quebec, ending the Revolutionary War and America's dream of independence.

Today, armies are no longer vanquished by smallpox, but concern over the safety of vaccines is on the rise. That concern is hardly new though and dates back to 1720 when inoculation was first developed. Even Benjamin Franklin distrusted the new procedure, tragically. During a smallpox epidemic in 1736, Franklin failed to inoculate his son who later died of the disease. Franklin blamed himself for his son's death later writing in his autobiography:

> "In 1736 I lost one of my sons, a fine boy of four years old, by the smallpox taken in the common way. I long regretted bitterly and still regret that I had not given it to him by inoculation. This I mention for the sake of the parents who omit that operation, on the supposition that they should never forgive themselves if a child died under it…"

Inoculation was dangerous, roughly 2 percent died from the procedure, a shocking number today, but far less than those dying of smallpox. So it's not surprising that when Dr. Edward Jenner developed the original smallpox vaccine in 1796 skeptics warned against the new vaccine, "to guard parents against suffering their children becoming victims to experiment," a claim familiar today. And like today, there were those who portrayed vaccines as a global menace wreaking havoc on the human race. An 1807 screed printed in London described vaccines as: "A mighty and horrible monster… [which] devores mankind—especially poor helpless infants… This monster has been named vaccination; and his progressive havoc among the human race, has been dreadful and most alarming."

The first organized anti-vaccine movement was the Anti-Vaccination League, created in England after the passage of the Vaccination Act of 1853. The Act made smallpox vaccinations compulsory for infants and held parents liable by fine, or even imprisonment, for failing to vaccinate their children. The unpopular act led to growing protests, culminating in 1885 when 100,000 angry protestors demanded the repeal of the act. After years of study, Parliament passed the Vaccination Act of

1898. The act introduced the concept of "conscientious objector" into English law, allowing parents who objected to vaccines on religious or other grounds to exempt their children from compulsory vaccination.

In 1879, the American Anti-Vaccination Society was formed after a visit by William Tebb, a British social reformer against compulsory vaccination. In 1893, Tebb published *Leprosy and Vaccination* in which he claimed that a recent increase in leprosy was due to smallpox vaccinations.

A century later, starting in 1998, similar claims would be made that vaccines led to autism. That year, the British physician Andrew Wakefield published a paper in the medical journal *The Lancet*. Dr. Wakefield's paper claimed that the measles, mumps, and rubella vaccine had contributed to the rise in childhood autism over the prior four decades.

Dr. Wakefield's claims were disproven after extensive research by multiple organizations consistently yielded no credible linkage between childhood vaccines and the claimed adverse effects. In 2010, the United Kingdom revoked Wakefield's medical license for fraudulent claims made in his 1998 paper, "citing, among other findings, that he had not disclosed funding from lawyers suing vaccine manufacturers."

But Andrew Wakefield's medical decertification did little to slow the anti-vaccine movement. For years, vaccine skepticism had been propelled by Robert F. Kennedy Jr., who, as early as 2005, had claimed vaccines caused autism.

Yet, every major public health organization including the Autism Society, Autism Science Foundation, and the National Institutes for Health had consistently confirmed that "Vaccines are Not Associated with Autism."

Vaccines, though, are not without risk. Current vaccine researchers are still cautioned by the 1955 Cutter Incident. Cutter Laboratories was a family-owned vaccine manufacturer. In the nation's first mass polio vaccination program, 200,000 children were administered Cutter's new polio vaccine. Within days, reports of paralysis began to appear. The vaccination program was quickly terminated, but not before 40,000 children had developed polio. Tragically, two-hundred children suffered permanent paralysis of varying degrees, and ten children died. Subsequent investigations determined that Cutter's process for inactivating the live polio virus had been defective.

But even accounting for that tragic accident, vaccines have saved countless more lives from polio deaths. In the 1940s, a virulent new polio strain doubled death rates. Fortunately, the new polio vaccines were a stunning success. One of medical science's greatest achievements, polio vaccines saved thousands of lives a year, and many more from paralysis and life in an "iron lung."

# The Great Barrington Declaration

MARCH 4, 2025

On October 4, 2020, three respected public health scientists—Drs. Jay Bhattacharya, Sunetra Gupta, and Martin Kulldorff—signed the Great Barrington Declaration. Weeks earlier, Dr. Scott Atlas had recruited the three scientists to help promote his COVID mitigation policies during his brief tenure on the White House COVID Task Force.

The Declaration was named for Great Barrington, Massachusetts, home to the American Institute for Economic Research where the three authors had signed the Declaration. Formed in 1933 during the Great Depression, the institute's stated mission is to "educate people on the value of personal freedom, free enterprise, property rights, limited government, and sound money."

The Declaration proposed the nation adopt "Focused Protection" allowing communities to reopen while protecting the elderly and otherwise vulnerable. The Declaration's authors believed that their approach would allow the COVID virus to harmlessly infect and subsequently immunize millions of healthy Americans. Once a significant majority of the population had been immunized, the nation would achieve "herd immunity" stopping the spread of the virus and ending the pandemic.

The Great Barrington Declaration quickly developed strong support including President Trump's endorsement. But within weeks, it became apparent that COVID death rates in many states that embraced the Declaration had begun to increase.

Iowa and New Hampshire were typical. Both states have similarly aged populations. Obesity rates for both states are near the 34 percent national average. During the first wave in early 2020, both Iowa and New Hampshire declared a public health emergency, including school closures. Both states had low COVID death rates.

Drs. Scott Atlas, Jay Bhattacharya, Sunetra Gupta, and Martin Kulldorff following a meeting with HHS Secretary Alex Azar. WHITE HOUSE

But during the second half of 2020, the two states adopted different approaches to managing COVID. Iowa chose to aggressively reopen while New Hampshire remained largely closed. After suffering similar death rates during the first half of the year, Iowa and New Hampshire deaths diverged in the second half. From July through December, New Hampshire's cumulative COVID death rate was 430 COVID deaths per million while Iowa's death rate soared to 1,000 deaths per million.

Iowa and New Hampshire were not unique. During the second half of 2020, the ten most open states—South Carolina, Nebraska, Iowa, Arkansas, North Dakota, Oklahoma, Wyoming, Utah, Wisconsin, and South Dakota—averaged twice the COVID death rates (930 deaths per million) relative to the ten most closed states: California, Colorado, Hawaii, New Jersey, New Mexico, New York, Maine, Maryland, Delaware, and Washington (430 deaths per million).

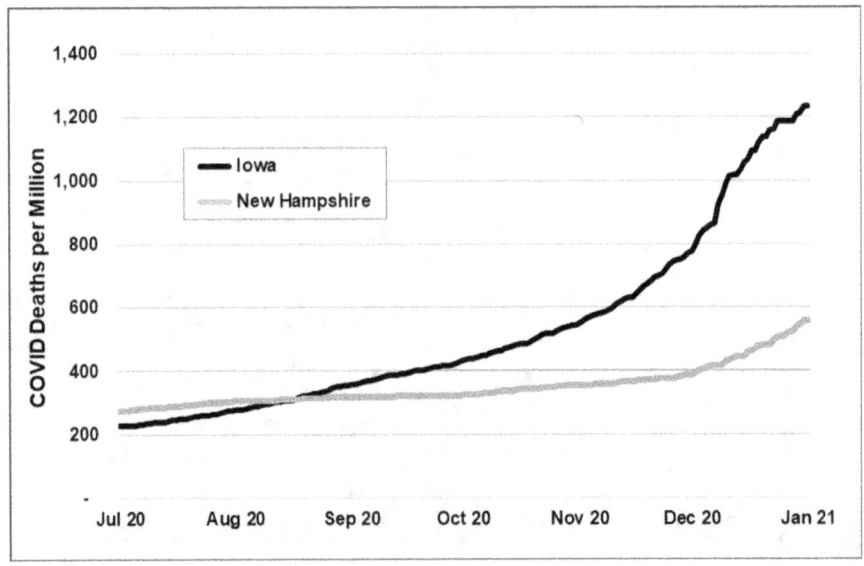

During 2020, COVID death rates were generally higher in states that aggressively reopened starting in the Summer. CDC

Although the Declaration's Focused Protection enjoyed wide support, many epidemiologists believed its reasoning was flawed.

At the heart of the controversy was the Declaration's claim that "vulnerability to death from COVID-19 is more than a thousand-fold higher in the old and infirm than the young. Indeed, for children, COVID-19 is less dangerous than many other harms, including influenza." That's true for the young, but for those between thirty-five and sixty-five years of age, their

COVID death rates, while less than the "old and infirm," were still significant.

In addition, the Declaration's authors seemed to have overlooked a critical detail: there are far more young and middle-age Americans than old and infirm. So even though the young and middle-aged died at lower rates during the pandemic, 270,000 Americans under the age of sixty-five died of COVID, a quarter of all COVID deaths—not to mention the 250,000 between the ages of sixty-five and seventy-four who also succumbed to the virus.

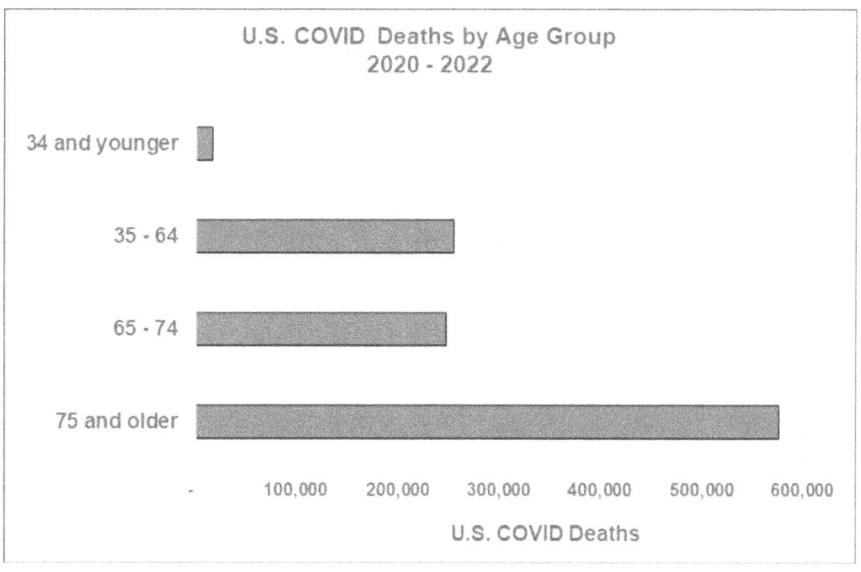

A quarter of COVID deaths were young and middle-aged. CDC

These deaths were tragic, cutting short many thousands of lives which could have been devoted to parenting the next generation, growing the nation, and nurturing society.

The Great Barrington Declaration divided America. Republican states embraced the Declaration, Democratic states rejected it. Even more divisive though were the COVID vaccines which became available in December 2020. The vaccines had been developed in record time under President Trump's

Operation Warp Speed, arguably his signature accomplishment during his first term. Yet, Republican states were slow to accept the new and as many believed, dangerous, vaccines.

Texas and Vermont are good examples of America's divided approach to the COVID vaccines. Both states share similar histories as proud, independent countries before joining the United States.

Vermont broke away from neighboring New York in 1777 as the American colonies were beginning their War of Independence. The Vermont Republic remained independent for fourteen years until 1791, when the Continental Congress welcomed Vermont as the fourteenth state. Today, the 1776 Green Mountain Boys flag is proudly flown by the Vermont National Guard.

Decades later, in 1836, Texas rebelled from Mexico and became the Republic of Texas before joining the Union in 1846 as the twenty-eighth state. Much of Texans' Independent heritage was forged during the Battle of the Alamo, fought during the Texas Revolution against Mexico.

Today, Texas and Vermont remain proudly and stubbornly independent. That stubborn independence was manifest during the pandemic. Texas prioritized personal liberty while Vermont focused on public health.

Midland County, Texas and Chittenden County, Vermont share many attributes. Prior to the pandemic, both had similar average lifespans. Both counties have median household incomes comfortably above the national average. Both counties are anchored by medium-sized cities, Midland, Texas and Burlington, Vermont.

The counties are also different. A third of Midland's population is obese compared to Chittenden's 20 percent. The Texas health care system is ranked forty-second in the nation. Vermont is ranked seventh. One in five adults in Midland

County are uninsured compared to one in twenty in Chittenden County.

The greatest difference between the two counties, though, is political affiliation. In 2020, Midland County cast 76 percent of its votes for Donald Trump while 75 percent of Chittenden County voted for Joe Biden.

Political orientation dictated the two states' public health policies during the pandemic. Texas remained defiantly open, while Vermont was the nation's most closed state. By late 2022, two years after the COVID vaccines had been released, only 44 percent of Midland's population had been fully vaccinated compared to Chittenden's 78 percent.

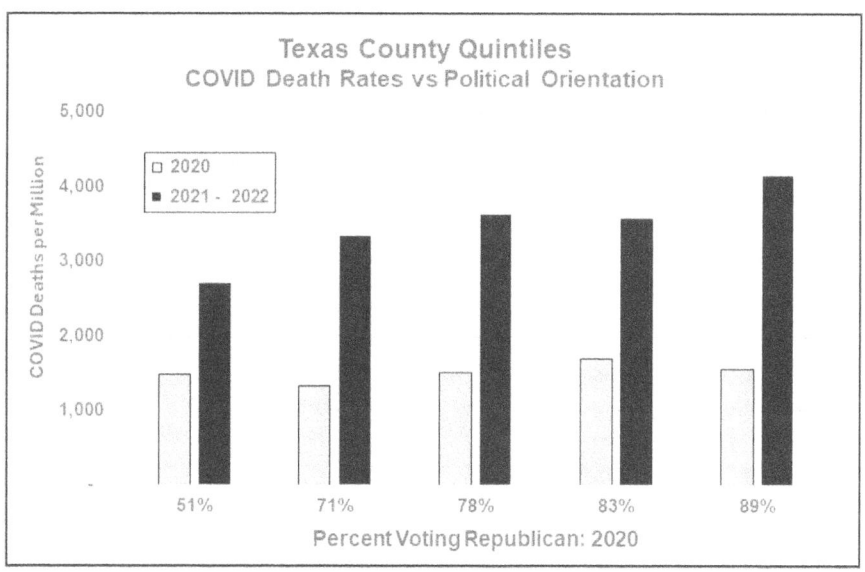

Vaccine hesitancy, largely among Republicans, cost thousands of lives.     CDC

Not surprisingly, the counties' different public health policies affected their COVID death rates. During 2021 and 2022, Midland County suffered 1,900 deaths per million while Chittenden County had only 800 deaths per million.

But just a few hundred miles east of Midland lies Austin, the Texas state capitol located in Travis county. A Democratic enclave inside Texas, Austin gave 71.4 percent of its vote to Joe Biden in 2020, and was 72 percent vaccinated by the end of 2022. Almost certainly, thanks to its high vaccination rate Travis county had only 990 COVID deaths per million during 2021 and 2022—half Midland county's death rate.

Similar to Midland and Travis counties in Texas, large differences in voting patterns, and COVID death rates, were common within many states. So a county-level analysis most clearly shows how political affiliation relates to public health policies.

The chart above plots COVID death rates. Deaths varied little during 2020 across the state's 251 counties; that was before vaccines were available.

But during 2021 and 2022, significant differences in death rates developed. The most Republican, and least vaccinated, counties had 55 percent more COVID deaths than the most vaccinated counties. The difference cost Texans thousands of avoidable deaths.

Texas was not unique. Across America, political orientation largely determined how states, counties, even cities responded to COVID.

# President Trump Declares COVID a National Emergency

MARCH 11, 2025

Five years ago today, the World Health Organization declared the COVID outbreak a global pandemic. Two days later President Trump declared a national emergency asking "...all Americans, including the young and healthy, work to engage in schooling from home when possible. Avoid gathering in groups of more than ten people. Avoid discretionary travel. And avoid eating and drinking at bars, restaurants, and public food courts."

The President's declaration marked the beginning of America's worst influenza pandemic in a century taking over one million American lives.

We Americans take pride in our American Exceptionalism. Yet, during the pandemic, the United States suffered more COVID deaths than during all its foreign wars, the largest employment loss since the Great Depression, and the greatest increase in the national debt since the Second World War. No developed nation had a worse record.

Understandably, we Americans want to move on. Yet, we owe it to ourselves, and to future generations, to understand why America stumbled so badly during one of our nation's greatest challenges.

I wrote COVID WARS to help answer that question.

# SHORT ESSAYS FOR INQUIRING MINDS

"Gruner's analysis of data is transparent and persuasive," a prominent reviewer wrote. "While he explores the history of virology and pandemics and other nations' responses to Covid, what is perhaps most powerful here is revisiting, in Gruner's precise and unheated reporting, the feeling of a nation spinning out of control, especially as overwhelming and contradictory conspiracy theories, accusations, and misinformation proliferated."

For more details scan the QR code:

https://www.amazon.com/dp/1737823160

# How President Clinton Let Russia Slip Away

MARCH 18, 2025

As President Clinton entered office, Russia was emerging from the once-powerful Union of Soviet Socialist Republics (USSR). A smaller, weaker, and humbled country, Russia renamed itself the Russian Federation.

Russia's economy had collapsed, with its GDP declining from $517 billion in 1990 to $196 billion at the end of the decade in 1999, only slightly larger than Indiana's gross state product. Military spending plummeted, falling to $6.5 billion by 1999, just two percent of America's military budget and on a par with the Netherlands.

With the end of the Cold War, the North American Treaty Organization (NATO) had lost its original mission. But rather than disband or even just reduce its footprint, a 1991 NATO strategy paper concluded that "the changed environment offers new opportunities for the Alliance to frame its strategy within a broad approach to security."

New opportunities. Like all bureaucracies, NATO was seeking not only to justify its existence, but also new opportunities to expand. NATO could have accepted Russia as a member, as it had Germany after World War II. But it did not. Should its erstwhile adversary become a part of NATO, and with no serious threats on the horizon, who exactly would NATO be protecting its members from?

The end of the Cold War represented the international organization's second missed opportunity to engage the Soviets. In 1954, the Soviet Union, a World War II ally of the United States, had applied for NATO membership.

But NATO rebuffed the Soviet request and instead, a year later, invited West Germany as a member. As part of its membership, West Germany was allowed to rebuild its military, a move Moscow considered highly threatening. The Soviets had hardly forgotten that a little over a decade earlier, Germany's invasion of Russia had resulted in the deaths of an estimated twenty-five million Russians.

Russian President Boris Yeltsin and President Bill Clinton during Yeltsin's 1995 visit to the United States. ALAMY

So, it was little surprise that two weeks after West Germany joined NATO, Russia and its six satellite states signed the Warsaw Pact as the Soviet Union's counterbalance to NATO, and the threat of a German resurgence.

NATO's 1954 rebuff of the Soviets was not the first time Western Europe had rejected Soviet overtures.

Russia's estrangement with the West can be traced back to April 16, 1939. That day, Soviet Premier Josef Stalin proposed forming a triple alliance that consisted of the USSR, Great Britain, and France to counter Germany's territorial ambitions. Stalin's proposal had a solid historical precedent: the three nations had been allies in World War I against Germany.

But Neville Chamberlain, Britain's Prime Minister, wasn't interested. Seven months earlier, Chamberlain and German Chancellor Adolf Hitler had signed the Munich Agreement in which Germany had promised to limit its territorial expansion. Chamberlain was confident that he had thwarted Hitler and achieved "peace for our time."

Winston Churchill disagreed and urged Chamberlain to accept Stalin's offer. "Ten or twelve days have already passed since the Russian offer was made..." Churchill warned in a May 4 Parliament speech:

> "There is no means of maintaining an eastern front against Nazi aggression without the active aid of Russia. Russian interests are deeply concerned in preventing Herr Hitler's designs on Eastern Europe. It should still be possible to range all the states and peoples from the Baltic to the Black Sea in one solid front against a new outrage of [German] invasion."

Chamberlain remained unmoved. After months of failed diplomatic efforts with Britain, the frustrated Soviets signaled to Berlin it was open to a proposal from Germany. Unlike the British, the Germans moved quickly.

Three weeks after their first meeting, on August 23, 1939, the Soviet Union and Germany signed a non-aggression pact. With the USSR now aligned with Germany, Hitler invaded Poland a week later, on September 1. World War II had begun.

One of history's great questions is what would have been the outcome if Chamberlain had agreed to Stalin's triple alliance proposal? Declassified Soviet documents released in 2008 document that the Soviets intended to commit "120 infantry divisions (each with some 19,000 troops), 16 cavalry divisions, 5,000 heavy artillery pieces, 9,500 tanks and up to 5,500 fighter aircraft and bombers on Germany's borders in the event of war in the west." Such a huge force on Germany's eastern front may well have deterred Hitler from invading Poland, the spark that led to World War II.

Unfortunately, as happened in 1939 and again in 1954, the West rebuffed Russia as the Cold War ended. Rather than embracing the struggling country—as the United States had done with Germany and Japan after the Second World War—NATO viewed the Soviet Union's collapse as a recruiting opportunity and welcomed the Czech Republic, Hungary and Poland as members in 1999.

Russia bitterly viewed NATO's recruitment of its former Warsaw Pact partners not only as a threat, but a broken promise made by the United States and its European allies. As part of the 1990 agreement to the reunification of East and West Germany, declassified documents strongly suggest that Gorbachev was promised NATO would not expand "as much as a thumb's width further to the east" of Germany.

President Clinton, though, didn't feel constrained by the promises of the Bush administration. A year after taking office, during a speech in Prague, Clinton asserted: "The question is no longer whether NATO will take on new members, but when and how."

America's most respected Cold War diplomat, George F. Kennan, was strongly opposed to NATO's eastward expansion. No U.S. diplomat understood the Soviet psyche better than Kennan. In 1946, while stationed in Moscow, Kennan had sent his famous Long Telegram to the State Department condemning the Soviet leadership and predicting the Soviet Union's expansion into Eastern Europe.

Now, with the Soviet Union defeated, Kennan argued against NATO expansion to the east, a move he believed would threaten Russia and reignite East-West tensions. Writing in *The New York Times*, Kennan, in a 1997 article predicted:

"Expanding NATO would be the most fateful error of American policy in the entire post-cold-war era. Such a decision may be expected to inflame the nationalistic, anti-Western and militaristic tendencies in Russian opinion; to have an adverse effect on the development of Russian democracy; to restore the atmosphere of the cold war to East-West relations, and to impel Russian foreign policy in directions decidedly not to our liking... And it is doubly unfortunate considering the total lack of any necessity for this move."

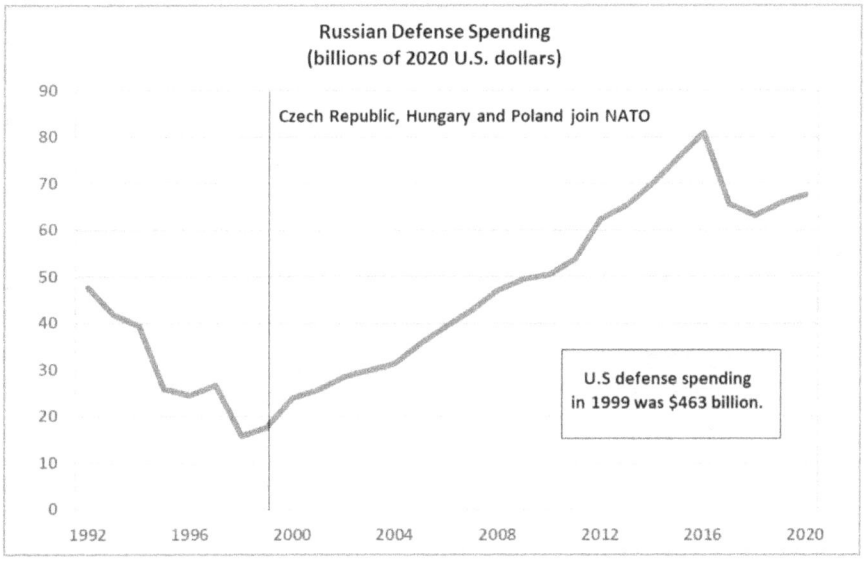

After declining for nearly a decade, Russian defense spending began to increase after three former Soviet Bloc countries joined NATO.
STOCKHOLM INTERNATIONAL PEACE RESEARCH INSTITUTE

Kennan's advice was ignored. On May 2, 1998, after the Senate had ratified NATO's eastward expansion, the ninety-four-year-old Kennan, in a *New York Times* interview, predicted:

> "I think it is the beginning of a new cold war. I think the Russians will gradually react quite adversely and it will affect their policies. I think it is a tragic mistake. There was no reason for this whatsoever. No one was threatening anybody else."

Kennan was right. As NATO began to absorb Russia's former allies, Russia quickly reversed its ten-year decline in defense spending and began to rebuild its military.

Years later, Sergei Karaganov, a prominent Russian political scientist, summarized Moscow's view of the "victors' peace" dictated by the West after the collapse of the Soviet Union:

> "The West has consistently sought to expand its zone of military, economic and political influence through NATO and the [European Union]. Russian interests and objections were flatly ignored. Russia was treated like a defeated power, though we did not see ourselves as defeated. A softer version of the Treaty of Versailles was imposed on the country."

President Truman's policies after World War II turned Germany and Japan into two of America's greatest allies. President Reagan's diplomacy helped to end the Cold War. President Bush had patiently overseen the collapse of the Soviet Bloc and the reunification of Germany. But President Clinton let Russia slip away.

# The Rise and Fall of the Department of Education

MARCH 25, 2025

Last week on March 20, President Trump issued an executive order promising to "take all necessary steps to facilitate the closure of the Department of Education and return authority over education to the States and local communities."

President Trump's decision to close the Department of Education isn't surprising, the department has been controversial since its formation in 1979 under President Carter. At the time, there were some 276 education-related programs spread across 24 federal agencies. So a consolidation would seem to have made sense.

During his 1976 presidential campaign, Carter had promised the National Education Association (NEA) — essentially a national teachers' union — that he would support the formation of an independent Department of Education. In return, the NEA gave Carter its first presidential endorsement in its then 117-year history.

But Carter had his doubts. Once becoming president, Carter slow-walked efforts to form the new department. "If we can work out some independent agency just for education where the teachers don't dominate it," Carter wrote in his diary on April 27, 1977, "then I would favor the idea." Hardly an enthusiastic endorsement.

SHORT ESSAYS FOR INQUIRING MINDS

With lukewarm support, it wasn't until October 1979 that Congress passed the Department of Education Act.

If Carter had his doubts about a Department of Education, President Reagan did not. During his first State of the Union speech, Reagan promised to dismantle the new Department of Education. Reagan believed that decisions about education should be made at the local level and that the federal government should play only a minor role in the nation's schools. But blocked by Congress, President Reagan was unable to close the new Department.

On March 20, 2025, President Trump signed an executive order to dismantle the Department of Education.  NEW YORK TIMES

For the next twenty years the Department of Education grew slowly until President George W. Bush took office in 2001. A year after assuming the presidency, Bush signed the No Child Left Behind Act (NCLB). In choosing to make education a primary focus of his presidency, President Bush was, no doubt, influenced by his wife and First Lady, Laura Bush. Early in her career, the First Lady had been an elementary school teacher and librarian.

The NCLB Act had three objectives. First, increase student learning especially in reading and mathematics. Second, reduce achievement gaps across the nation's student population. Third, increase accountability for schools, teachers, and states through regular, standardized testing.

The NCLB Act was expensive. During the eight years of the Clinton presidency, Department of Education annual spending averaged $30.2 billion with little annual growth. Under President Bush, education spending doubled to $62.5 billion annually.

All that extra spending achieved little. Ten years after the NCLB Act was enacted, student learning, as measured by nation-wide tests, had improved only slightly. In 2002, the average state score for fourth grade reading, for example, was 217. By 2019, the average score had crept up to 219, a hardly measurable increase.

Perhaps the most significant result (for better or worse based on one's viewpoint) of the No Child Left Behind Act was the light the Act put on the large achievement gaps among low-income, minority, and disabled students. It was this realization that largely led to the emphasis on Diversity, Equity, and Inclusion (DEI) programs across the nation's schools, businesses, and institutions over the last two decades. These programs were largely terminated by President Trump after he resumed his presidency two months ago.

During the COVID pandemic, the Department of Education, public health agencies, and teachers' unions were often blamed for unnecessarily closing schools to the detriment of school children. The lockdowns during the pandemic disrupted lives, but school lockdowns had less impact on students academically than generally believed.

For example, fourth grade reading test scores for the seven states that never closed their schools — Arkansas, Iowa, Nebraska, North Dakota, South Dakota, Utah, and Wyoming — declined an average 3.0 percent from 2019 to 2024. Surprisingly, the most closed states — Hawaii, Maryland, New Mexico, Oregon, Virginia, Washington, and California — declined slightly less, only 2.9 percent. COVID itself was as responsible for societal disruption as were the lockdowns. (For more discussion on the societal impact of lockdowns during the pandemic, see "Chapter 8: Lockdowns" in my book, *COVID WARS*.)

From the beginning, there have been large differences in test scores across the fifty states. These differences have changed little over the years. For example, in 2009 eighth grade reading scores varied from 274 in Massachusetts to 251 in Mississippi. Ten years later, in 2019, the scores ranged from 273 in Massachusetts to 252 in New Mexico.

What factors account for these large, persistent differences across the nation's fifty states? Certainly, school funding, teacher quality, and socioeconomic factors play a role. But one factor correlates very closely with student performance: the percentage of college graduates within the state. College graduates earn more, pay more taxes, commit fewer crimes, have fewer deaths of despair, have more stable families, and arguably place a higher value on education. All this contributes to better schools, more motivated students, and better grades.

Education improves nearly all aspects of life. The ten most highly educated states have average lifespans of eighty years versus seventy-six for the least educated, median household incomes (2019) of $87,600 versus $60,100, and during the pandemic suffered 2,800 COVID deaths per million versus 4,100 per million (and were 78 percent fully vaccinated versus 59 percent).

The chart below illustrates the close correlation of state education levels, measured by the percentage of college graduates within the state, to the combined reading and mathematics test scores for eighth graders.

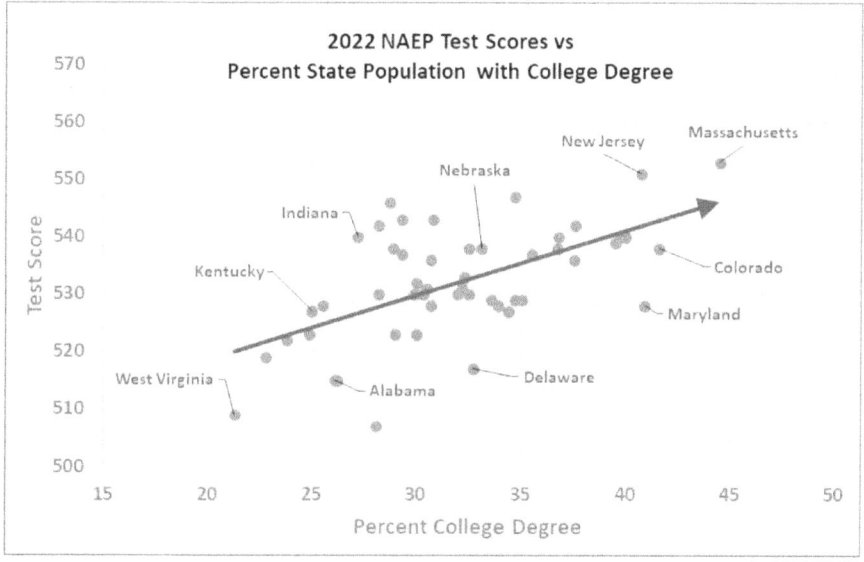

Across the nation, a state's student test scores typically correlate with the state's percentage of college graduates.  NAEP | FRED

Although President Bush's No Child Left Behind initiative was unsuccessful, it was a noble and worthy effort. Returning control of education back to the states is consistent with the American Constitution. The Constitution never mentions education, and declares in the Tenth Amendment, "The powers not delegated to the United States by the Constitution, nor prohibited by it to the States, are reserved to the States respectively, or to the people."

So, states will, once again, be responsible for their children's education. Their success will depend on their values and the subsequent priority they place on education. On December 26, 2024, former presidential candidate Vivek Ramaswamy generated a political firestorm when he tweeted:

> "Our American culture has venerated mediocrity over excellence for way too long (at least since the 90s and likely longer). That doesn't start in college, it starts YOUNG. A culture that celebrates the prom queen over the math olympiad champ, or the jock over the valedictorian, will not produce the best engineers."

Ramaswamy's emphasis on academics over athletics may be right (if the objective is to produce the best engineers). In 2022, the twenty-five states with the highest percentage of college graduates averaged 540 on the above NAEP tests while the lower twenty-five states averaged only 524.

During the March 20 ceremony signing the executive order to dismantle the Department of Education, President Trump declared:

> "We're at the bottom of the list and we're the most expensive. We're at the top of the list when it comes to cost per pupil. We spend more money per pupil than any other nation in the world and yet we're rated No. 40. The last ratings came out, you saw them. They talk about 40 countries. We're rated No. 40."

The President's claim that the United States spends more per student on education than any other country was largely right, only four countries spend more. In 2019, the United States spent $15,500 per elementary and secondary school student, noticeably higher than other developed countries including Australia ($14,100), Sweden ($13,800), Germany ($13,700), United Kingdom ($13,300), Canada ($12,800), and France ($12,200).

America's high level of educational spending has not resulted in better test results, or even scores comparable to most other developed nations. Every three years, fifteen-year-old students from approximately eighty countries take the Programme for International Student Assessment (PISA) test.

The PISA test measures attainment in reading, mathematics, and science. As the table below shows, in 2018 American students tested poorly relative to other developed countries.

**PISA Test Scores**
**Total Scores: Reading, Mathematics, and Science**
**Ranked by Score**

| 2018 | | 2022 | |
|---|---|---|---|
| China | 1736 | China | 1605 |
| Japan | 1560 | Japan | 1599 |
| South Korea | 1559 | South Korea | 1570 |
| Canada | 1550 | Canada | 1519 |
| Finland | 1549 | Ireland | 1512 |
| Ireland | 1514 | Australia | 1492 |
| United Kingdom | 1511 | Finland | 1485 |
| New Zealand | 1508 | New Zealand | 1484 |
| Sweden | 1507 | United Kingdom | 1483 |
| Netherlands | 1507 | Denmark | 1472 |
| Denmark | 1503 | **United States** | **1468** |
| Germany | 1501 | Sweden | 1463 |
| Belgium | 1500 | Belgium | 1459 |
| Australia | 1497 | Austria | 1458 |
| Norway | 1490 | Germany | 1447 |
| **United States** | **1485** | Netherlands | 1440 |
| France | 1481 | France | 1435 |
| Austria | 1473 | Italy | 1430 |
| Italy | 1431 | Norway | 1423 |
| Israel | 1395 | Israel | 1397 |

Student test scores declined in nearly all countries during the COVID pandemic. But the average decline in U.S. test scores was smaller than most other countries.

PROGRAMME FOR INTERNATIONAL STUDENT ASSESSMENT

# A Trade War over Chickens Led to America's Love for Pickup Trucks

APRIL 1, 2025

We don't know for sure, but we're guessing that in 1964 Frank Perdue, the chicken magnate, was pretty darn mad. Since 1939 when he dropped out of college, Frank had been working at his family business helping to make chicken dinners a mainstay of American cuisine. He and his arch competitor, Don Tyson, had borrowed Detroit's approach to mass production to turn chicken farming from a mom and pop business into a major industry.

It was hard work breeding, raising, and processing billions of chickens a year just so Americans could enjoy their fried chicken dinner after church on Sunday. But Frank was more than up to the task. As he liked to say, "It takes a tough man to make tender chicken."

By the early nineteen-sixties, Frank and his competitors were looking for broader markets. But Europe, and particularly Germany, weren't interested. They had their own chicken farmers, typically small family farms selling chickens to markets that still considered chicken something of a delicacy. To keep American chickens out of their markets, by 1962 European regulators had raised tariffs on American chickens to nearly 50 percent.

America had to retaliate, but how? European farm exports into the U.S. were small. Taxing them, even with large tariffs, would have little impact. But the German automobile industry was a different matter. From Volkswagens to Mercedes, the Germans were making deep inroads into the American automobile market.

And then there was the VW Minibus. Beloved as cheap transportation by America's growing Hippie communes, the Minibus was a symbol not only of youthful rebellion, but Germany's increasing threat to the American automobile industry.

The most popular American and European automobiles in 2024 could not be more different.

So, President Johnson chose to retaliate by placing a 25 percent tariff on all imported "light trucks" which included the subversive VW Minibus, but also all pickup trucks — then driven primarily by farmers, ranchers, and construction workers. Decades later, most Sport Utility Vehicles (SUVs) would also fall under the light truck designation protecting these popular new vehicles from foreign competition.

President Johnson's decision to target imported light trucks was largely influenced by Walter Reuther, the president of the United Auto Workers labor union. But we're guessing, just guessing, that Johnson was also happy to stick it to those long-haired hippies.

Sixty-years later, President Johnson's Chicken Tax is still in effect providing a generous 25 percent price umbrella for

American manufacturers of pickups and large SUVs. These high tariffs have fundamentally altered the American car market and, in particular, America's ability to sell cars into global markets starting with Europe.

In 2024, according to *Car and Driver* magazine, four of the top five vehicles sold by American car makers were pickups, the innovative Tesla Model Y being the sole exception. The average price for the five vehicles in 2024 was $58,800.

The Chicken Tax hugely influenced American car manufacturers. Shielded from foreign competition, the American pickup has evolved from a pedestrian work vehicle to a symbol of America itself: powerful, rugged, and confident. And profitable. Industry analysts estimate that the Ford F-series pickups contribute 90 percent of Ford's global profits.

After the top five vehicles — four pickups and a Tesla — only four of the next twenty most popular vehicles purchased in the United States were made by American companies: the Jeep Grand Cherokee (ranking 12th), Chevrolet Equinox (13th), Chevrolet Trax (15th), and Ford Explorer (16th). All are SUVs and likely benefit from the Chicken Tax price umbrella. Outside of these four American vehicles, Toyota, Honda, and Nissan along with a sprinkling of Subarus and Kias dominate the American automobile market.

The European automobile market is drastically different. In 2024, according to the Motor1.com website, Europe's top five vehicles would all be classified as a compact or smaller by American standards. Their average price was €29,800, or about $32,200. The top selling vehicle was the Dacia Sandero manufactured by Renault's Dacia subsidiary in Romania. Tesla just missed the top five; its Model Y came in sixth.

For years, American automobile companies attempted to compete in Europe: General Motors with Vauxhall and Opel, Ford with European versions of its Escort, Granada, Fiesta, and Focus. But although there have been successes — the Ford

Focus was the European Car of the Year in 1999 and a runaway best-seller — American car executives never seemed to have their heart in selling small cars.

In 1976, when Washington issued the first gasoline mileage standards in response to the 1973 Oil Embargo, General Motors President E.M. Estes claimed the new standards would reduce the automobile market in which "the largest car the industry will be selling in any volume at all will probably be smaller, lighter and less powerful than today's compact Chevy Nova."

Estes' protests were on the mark when he lamented fuel economy would result in smaller cars like the Chevy Nova — in his mind a poor excuse for a car compared to the luxurious, much larger Chevy Impala. For the next fifty years, the world's best-selling car would be the Toyota Corolla. The Corolla weighed 2,885 pounds, far less than Estes' scorned Chevy Nova. (For a discussion of how the United States lost the global automobile market, see the Reagan chapter in my book, *We The Presidents*.)

Today, it's hard to imagine Detroit's auto executives, spoiled by their high profits manufacturing giant pickups, would be interested in competing in Europe where the best selling car is a cheap sub-compact manufactured in Romania.

Sadly, most Americans aren't buying American cars, but instead prefer Japanese and South Korean automobiles. Why would Europeans (or Asians and South Americans) feel any differently? If Europeans wished to import more cars, they would look to Asia, not North America.

President Trump intends to announce his global tariff plan on April 2, a day Trump has declared to be "Liberation Day." The President has promised that his tariff plan will usher in an American Golden Age. "For DECADES," President Trump posted on March 21, "we have been ripped off and abused by every nation in the World, both friend and foe. Now it is finally time for the Good Ol' USA to get some of that MONEY and RESPECT, BACK. GOD BLESS AMERICA!!!"

Forty years ago, President Reagan also tried to stop the influx of imported cars — smaller, more fuel efficient, and higher quality than their American counterparts — into the United States. Reagan used import quotas to limit Japanese imports since tariffs had been, for decades, anathema to conservative Republicans.

Reagan failed. The demand for Toyotas, Hondas, and Nissans was so high that Japanese profits skyrocketed after the Japanese manufacturers raised their prices to manage the strong demand. The manufacturers then used their profits to develop the Lexus, Acura, and Infiniti luxury brands to compete with American luxury cars.

Unlike President Trump, Reagan never blamed others for America's trade imbalance. "The way up and out of the trade deficit," President Reagan explained during a May 1987 radio address, "is not protectionism, not bringing down the competition, but instead the answer lies in improving our products..."

But for the next three decades the United States failed to produce an automobile that could compete on the global stage — until Elon Musk introduced his revolutionary all-electric Tesla.

Few Americans realize that the best-selling automobile worldwide in both 2023 and 2024 was American: the Tesla Model Y.

Yet Tesla has been mired in controversy. From its inception, conservative Americans despised Tesla, and their owners, for being "Woke." For them, true Americans drove cars fueled by American oil and cheered when Trump called for America to "Drill, Baby, Drill!"

As 2023 was coming to a close, and Tesla had regained for the United States the mantle of global leadership in automobiles, Donald Trump posted his Christmas message on Truth Social. During the message, Trump lumped the "All Electric Car Lunacy" in with his political enemies that "are looking to destroy our once great USA. MAY THEY ROT IN HELL." Such was the support from the former and future American President for the first American car in living memory to be the world's best seller.

After partnering with Elon Musk, President Trump now loves Teslas, even promoting them on the White House lawn on March 11 when he bought a Tesla Model S to be used by White House guests.

But the remarkable genius behind Tesla, Elon Musk, seems to have come unhinged after he aligned with Trump starting in July 2024. By March 2025, Musk, as the unofficial head of DOGE (Department of Government Efficiency), was rampaging through the federal government after promising to save two trillion dollars in government waste, fraud, and abuse.

Today, after opining on European politics and accused of mimicking a Nazi salute, much of Europe is rejecting Elon Musk and the Tesla brand. Here in America, liberal Americans are selling their Teslas and savaging Tesla dealerships.

Meanwhile, Elon Musk is in Wisconsin giving away million dollar checks hoping to encourage the state's voters to elect a conservative Supreme Court judge. "I think this will be important for the future of civilization," Musk said. "It's that significant."

# Looted, Pillaged, Raped, and Plundered

APRIL 8, 2025

In 1962, Milton Friedman, one of the last century's most prominent economists, wrote a landmark book, *Capitalism and Freedom*. For eight years, Friedman's book was just another academic tome gathering dust on professors' bookshelves. But in 1970, Friedman summarized his book in a legendary *New York Times* magazine article. Friedman's message was simple:

> "There is one and only one social responsibility of business—to use its resources and engage in activities designed to increase its profits."

Friedman's economic philosophy gave corporations the moral justification to focus on maximizing profits over all other considerations, their only limitation being to engage in "open and free competition without deception or fraud."

Friedman went a step further. Not only should corporations focus solely on profits, their "use of the cloak of social responsibility, and the nonsense spoken in its name by influential and prestigious businessmen, does clearly harm the foundations of a free society."

In short, corporations must fully reject "the cloak of social responsibility." Individuals might strive to be good citizens, but responsible citizenship, according to Friedman, is forbidden to corporations whose sole responsibility is to their shareholders.

America's corporations quickly embraced Friedman's philosophy. Profit maximization became such accepted corporate dogma that by the late nineteen-seventies the CEO of Burroughs, a major computer manufacturer at the time, could proudly boast that Burroughs' tight-fisted corporate spending kept their locked-in customers "sullen, but not rebellious." (This was during a period when computer software was proprietary, making it very difficult for users to change computer manufacturers.)

Economist Milton Friedman and General Electric CEO Jack Welch.
WIKIPEDIA

The singular focus on corporate profits propelled the upward shift in American wealth. America's middle-class was built in the postwar years from 1946 through 1972 when real, after-tax incomes increased from $25,800 (in 2020 dollars) to $44,400, an average annual increase of 2.2 percent. But after 1972, as Friedman's philosophy of profit maximization became entrenched, middle-class incomes began a slow decline, falling to $43,000 by 2018.

The affluent (the top income quintile), on the other hand, thrived and enjoyed an increase in after-tax income from $118,500 in 1972 to $201,600 in 2018. The truly wealthy, the top one percent, did far better, growing their after-tax incomes in those same years from $333,000 to $1,193,000. Much of that new wealth came from stock appreciation and dividends driven by rising corporate profits.

No American businessman mastered the Friedman philosophy better than General Electric's CEO, Jack Welch. In 1999, *Fortune* magazine proclaimed Welch "Manager of the Century," after growing GE's profits and stock price to stratospheric levels.

A brilliant but ruthless manager, Welch explained his global business strategy to Lou Dobbs on CNN Moneyline. "We've never had a better opportunity to source joint ventures around the globe… Ideally, you'd have every plant that you own on a barge to move with currencies and changes in the economy."

The American business community embraced Welch's hard-nosed strategy which Businessweek summarized as:

GE's U.S. workforce has been shrinking for more than a decade as Welch has cut costs by shifting production and investment to lower-wage countries. Since 1986, the domestic workforce has plunged by nearly 50 percent, to 163,000, while foreign employment has nearly doubled, to 130,000… GE has made clear its desire that its suppliers [also] move to Mexico.

For globe-straddling, multinational companies, labor had become a mere commodity to be acquired at the lowest possible cost. And why not? One of America's most iconic companies, General Electric, felt no loyalty to American workers. In GE's ideal world, factories would be located on barges to move freely around the globe seeking cheap labor.

President Trump announced his new tariff plan on April 2, or "Liberation Day" as he dubbed his highly promoted tariff announcement. During his speech, Trump claimed that for decades scores of greedy countries had "looted, pillaged, raped, and plundered" our nation costing Americans millions of jobs and trillions of dollars.

To reverse these huge losses, President Trump announced a series of "reciprocal tariffs" designed to reduce America's large trade deficit. The President proposed that the reciprocal tariff for each country be based not on their actual tariff rates but on their trade deficit with the United States. If a nation had a large U.S. trade deficit, America would reciprocate with large tariffs.

The White House dressed-up the calculation of reciprocal tariffs in their official release using Greek symbols and two superfluous elasticity variables which together null out.

Greek letters aside, the calculation of the President's reciprocal tariff is simple. The new tariff is calculated by: (1) subtracting the nation's imports from its exports, (2) dividing the difference by the imports, and (3) dividing the result by two. The minimum tariff is 10 percent, even for countries with positive U.S. trade balances.

For an example, consider Vietnam. From March 2024 through February 2025 Vietnam exported $142.34 billion in goods to the United States while importing $13.38 billion. Based on this trade data, Vietnam's reciprocal tariff would be:

$$\text{Reciprocal Tariff} = [(142.34 - 13.38) / 142.34] / 2$$
$$= 90.6 / 2 \approx 46\%$$

America's new 46 percent reciprocal tariff on Vietnam is far more than the tariffs Vietnam has historically charged the United States, typically 15 percent or less.

President Trump is right that the United States has large trade imbalances with many countries. Vietnam has one of the largest, and representative of many countries with which the United States has unfavorable trade balances.

For every dollar of goods that Vietnam imports from the United States, it exports nearly eleven dollars in goods to the U.S. That's a huge difference. But rather than "ripping off" the United States, Vietnam's large trade imbalance with the U.S. is largely due to three economic factors:

1. Vietnam's exports to the U.S. have grown from $49.1 billion in 2018 to $136.6 billion in 2024. This increase was primarily due to factories migrating from China to Vietnam after President Trump, during his first term,

pressured China to reduce its exports to the United States.
2. Vietnam is a relatively poor country. Its citizens are primarily concerned with meeting their basic needs for food, clothing, shelter, and transportation. America's exports, with the exception of certain food and mineral oil products, do not fulfill these basic needs.
3. As Milton Friedman proclaimed, corporations have no "social responsibility" for trade imbalances, or anything else. Their sole responsibility is to maximize profits. That's done by purchasing goods from low-cost countries regardless of the implications to America.

Although Vietnam has a huge trade imbalance with the United States, the reasons are not unique. The United States is a huge market for goods Americans no longer manufacture themselves from clothing to computers. This transformation was largely due to American companies abandoning American workers for cheap foreign labor.

As factories closed, the American economy shifted from production to consumption. Today, 81 percent of the U.S. economy, as measured by Gross Domestic Product (GDP), is devoted to consumption; few major countries spend more on themselves. Germany spends 74 percent, Switzerland 63 percent, Norway 59 percent, and China just 56 percent.

Countries with low consumption rates can devote more of their economy to the production of export goods. So it's not surprising that countries with low consumption rates like China, Germany, and Vietnam typically import less, and consequently have positive trade balances.

In contrast to President Trump's claims that countries have been "ripping off" the United States for decades, much of America's trade imbalance is due to the low consumption rates of many countries. They simply can't afford, or don't need what America is selling.

# The Life and Times of William McKinley

APRIL 15, 2025

William McKinley was a good man. Just weeks after the start of the Civil War, McKinley enlisted in the Union Army. He was eighteen years old.

For the next four years McKinley fought bravely at the Battle of Antietam in 1862, the Battle of Buffington Island in 1863, and multiple skirmishes throughout 1864. Steadily promoted during the war to the rank of captain, McKinley left the army as a brevet major, an honorary commission awarded for gallantry or meritorious conduct.

Five Civil War veterans would later serve as American presidents: Ulysses S. Grant, Rutherford B. Hayes, James A. Garfield, Benjamin Harrison, and William McKinley. All entered the war as officers, and left as generals, except one: William McKinley. McKinley joined the Army as an enlisted man, fought bravely—even after his horse was shot out from under him—and left the Army a war hero.

An outstanding soldier, McKinley's commanding officers encouraged him to join the peacetime army, but McKinley chose to study law. In 1867, McKinley opened a small law office in Canton, Ohio where he would reside, except for his years in Washington, D.C., for the rest of his life.

That same year, McKinley took his first tentative steps into politics when he campaigned for Rutherford B. Hayes who was running for governor in Ohio. During the war, Major Hayes had been one of McKinley's commanding officers and would remain a mentor and friend to McKinley until Hayes' death in 1893.

On January 25, 1871, McKinley married Ida Saxton, the popular and accomplished daughter of a prominent Canton family. The two met at a church picnic. The wedding of the upcoming young lawyer and the belle of Canton was attended by 1,000 guests at Canton's First Presbyterian Church.

President McKinley's second term campaign poster.　　WIKIMEDIA

William and Ida had a deeply loving, but tragic marriage. Their first daughter, Katie, was born Christmas Day, 1871. Katie was adored by her parents who commissioned both photo and oil portraits of their beloved daughter. A second daughter, named Ida after her mother, was born in the spring of 1873. Ida was a sickly child. She died four months later of cholera. Her mother was heartbroken and, fearful of losing her first daughter, never let Katie out of her sight except to take walks with the child's father.

Little Katie died in June 1875 of typhoid fever.

Ida McKinley never recovered from the death of her two daughters. Shortly after Katie's death, Ida developed epilepsy and would become medically and emotionally dependent on her husband. William was a devoted husband and cared for Ida the rest of his life.

In 1875, William McKinley had a thriving legal practice, had dabbled in local politics as the county's elected prosecuting attorney, and continued to support his wartime friend, Rutherford B. Hayes, as the Ohio governor.

The next year, in 1876, Hayes was nominated by the Republican party as their party's presidential candidate. With Hayes' encouragement, McKinley was mulling a run for office himself as Ohio's 17th District Congressional Representative.

But McKinley wasn't sure, concerned about the toll moving to Washington, D.C. would have on his frail wife. Ida insisted though that William pursue his political ambitions. For the next twenty-five years, Ida and William McKinley would be inseparable; Ida as a supportive, if fragile, political wife and William as a loving and devoted husband.

In 1877, both Hayes and McKinley won their elections, Hayes as president and McKinley as a Congressman. In a political compromise, Hayes had promised to serve a single term as president. Hayes kept his promise and after four years of a relatively undistinguished presidency, retired to his home in Ohio.

McKinley, though, spent twelve years in Congress quickly acquiring a powerful role on the House Ways and Means Committee and in 1889 its chairmanship.

From his first days in Congress, McKinley had supported tariffs designed to protect American industries. McKinley's motivations were understandable. By 1890, Canton, Ohio had become a major center for the manufacture of farm equipment. One manufacturer, the Aultman Company, had 10.5 acres under roof and employed 550 employees. With a constituency that was manufacturing-based, McKinley understandably supported protective tariffs.

McKinley's support of tariffs wasn't uncommon. Since the beginning of the Civil War, the federal government had depended on import tariffs as its primary revenue source. But by the 1880s, the government was running large surpluses, surpluses which could be better used by the private sector to finance industries and agriculture.

Congressional Democrats promoted lower tariffs both to reduce fiscal surpluses and consumer prices. Republicans argued for higher tariffs to provide protection primarily for the nation's businesses.

The Republicans, led by William McKinley, prevailed. The Tariff Act of 1890, dubbed the "McKinley Tariff," raised average import tariffs from 38 percent to 49.5 percent, the largest increase since the Civil War.

The backlash was immediate. Americans became angered as prices soared. Within a year, eighty-three Republicans, including William McKinley, had lost their seats in the House. That was just the beginning. In 1892, voters handed Democrats the presidency and both houses of Congress. Two years later, Congress passed the Wilson-Gorman Tariff Act of 1894 lowering tariff rates and eliminating tariffs on iron ore, coal, lumber, and wool.

McKinley, though, landed on his feet. Backed by Mark Hanna, a wealthy Cleveland businessman and political kingmaker, McKinley won the Ohio governorship in 1891 and again in 1893. But Mark Hanna had more ambitious plans for McKinley.

The years leading up to the 1896 presidential election were tumultuous. America was expanding economically as new industries built around railroads, steel, and oil were making a handful of capitalists, particularly Andrew Carnegie, John

Rockefeller, Cornelius Vanderbilt, and J.P. Morgan, fabulously rich.

But little of that wealth was trickling down to workers. The Homestead Strike of 1892 and the Pullman Strike of 1894, two of the largest labor strikes in American history, cost dozens of lives as thousands of striking steel and railroad workers, protesting for better pay and working conditions, battled with soldiers and police.

After years of rapid growth and speculation, a bank panic—the Panic of 1893—triggered a nationwide depression bankrupting thousands of banks, railroads, factories, and businesses. Unemployment rates soared to over 20 percent in parts of the country. It was the worst economic crisis until the Great Depression during the 1930s.

The 1896 presidential election revolved around two issues: tariffs and the gold standard. The high tariffs after 1890 had primarily benefitted industry and its owners by shielding American manufacturers from foreign competition. The gold standard though hurt workers and farmers. Backing the dollar with gold, rather than more abundant silver, limited the money supply resulting in deflation reducing wages and farm prices.

William McKinley, a self-proclaimed "tariff man," easily won the Republican nomination for president with Mark Hanna's financial and political support. The Democratic contender, thirty-six-year-old William Jennings Bryan, won the Democratic nomination with a single, electrifying speech, his famous Cross of Gold speech—a speech which over a century later still echoes today.

"There are two ideas of government. There are those who believe that if you just legislate to make the well-to-do prosperous, that their prosperity will leak through on those below. The Democratic idea has been that if you legislate to make the masses prosperous their prosperity will find its way up and through every class that rests upon it...

"If [Republicans] dare to come out in the open field and defend the gold standard as a good thing, we shall fight them to the uttermost, having behind us the producing masses of the nation and the world... We shall answer their demands for a gold standard by saying to them, you shall not press down upon the brow of labor this crown of thorns. You shall not crucify mankind upon a cross of gold."

McKinley won the presidential election sweeping up 271 electoral votes to Bryan's 176 votes. Just as America today is divided into red and blue states, America in 1896 was deeply divided. McKinley won the industrial states, Bryan won the agrarian states.

President McKinley quickly restored protective tariffs with the passage of the Dingley Tariff Act of 1897. The Act increased tariffs to an average of 52 percent, the highest protective tariff in the history of the United States.

In 1898, after being goaded by William Randolph Hearst's warmongering newspapers, the United States entered into the Spanish-American War. The war ended with Spain renouncing all rights to Cuba, ceding Puerto Rico and Guam, and selling the Philippine Islands to the United States for $20 million (approximately $770 million today). That same year, the United States annexed Hawaii after Queen Liliuokalani was overthrown in a rebellion encouraged by American sugar and pineapple interests. (Ambivalent in its new role as an imperialist nation, the United States granted Cuba its freedom in 1902, and the Philippines theirs after the Second World War in 1946.)

President McKinley won a second presidential term in 1900 with Theodore Roosevelt as vice-president. He would serve just six months.

McKinley's last major act as president was the the Gold Standard Act of 1900 firmly establishing gold, rather than silver, as the only metallic standard backing the U.S. dollar.

On September 6, 1901, Leon Czolgosz, a former steel working turned anarchist, shot President McKinley as the president was greeting guests at the Pan-American Exposition in Buffalo, New York.

As the angry crowd descended on Czolgosz, the fallen president implored the attackers to "Go easy on him, boys," and moments later, asked his Secret Service agents to "Be careful how you tell my wife."

For eight days, Ida McKinley remained at her husband's side comforting him as he lapsed in and out of consciousness. William McKinley died on September 14. For the next six years, Ida reportedly visited her husband's final resting place daily until her own death on May 26, 1907.

# Tariffs, Gold, Silver, and the Wizard

APRIL 22, 2025

During his second inaugural address, President Trump expressed his admiration for William McKinley declaring, "President McKinley made our country very rich through tariffs and through talent." Trump was right that McKinley had promoted protective tariffs for decades. But as I discussed in last week's Substack, McKinley's Tariff Act of 1890 was not only unpopular, but drove America into a deep recession costing Republicans both houses of Congress and the presidency in 1892.

But in a remarkable turn-around, William McKinley won the presidency in 1896 backed by Mark Hanna, a wealthy Ohio businessman and the nation's top political kingmaker at the time.

Late in his presidency, McKinley realized that tariffs had evolved from simply a source of government revenue, to a protective wall surrounding American industries, to a tool that could promote trade reciprocity across nations.

But first, a brief history of tariffs in America.

In 1791, Congress passed an excise tax on whiskey, the first major federal tax imposed on a domestic product. Rebellion quickly ensued. Americans were furious that their new government was taxing them just as the British had taxed the Colonists prior to the American Revolution.

The whiskey tax was especially unpopular in western Pennsylvania where whiskey production was a major source of income for the local farmers. The Pennsylvania farmers refused to pay the tax. Then in 1794 an angry mob burned down the house and tarred and feathered the local tax collector. Faced with rebellion, President Washington dispatched a militia force of 12,000 men to western Pennsylvania. The rebellion quickly collapsed.

Some historians believe that *The Wizard of Oz* is an allegory for the political issues during the McKinley presidency.
ALLSTAR | MGM | SPORTSPHOTO LTD

Congress had learned its lesson. For the next century, the federal government relied on easy-to-collect import tariffs as its primary source of revenue.

In 1861, Congress passed the Morrill Tariff Act. The Act's primary purpose was to protect America's growing industries from foreign competition. The Morrill tariffs increased the average tariff rate from 20 percent to 36 percent.

The new tariffs divided the nation. The industrialized northern states supported the protective tariffs while southern states, with little industry and reliant on foreign goods, detested the taxes.

From his first years as an Ohio Congressman, McKinley had been a strong supporter of protective tariffs. That's not surprising. Blessed with natural resources, populated by skilled workers, and crisscrossed with railroads extending throughout America, Ohio played a leading role in America's industrialization.

By 1900, the United States had become an industrial colossus with an estimated Gross Domestic Product (GDP) of $24 billion dwarfing the United Kingdom ($12 billion), Germany ($10 billion), and France ($7 billion).

High protective tariffs were no longer needed and were even damaging to American exports; nations were understandably hesitant to purchase American goods if America was unwilling to purchase theirs.

McKinley recognized that protective tariffs, once useful, were now damaging U.S. foreign trade. In a speech given at the Pan-American Exhibition in Buffalo, New York, the day before he was fatally shot by an assassin, McKinley stated:

> "Our capacity to produce has developed so enormously and our products have so multiplied that the problem of more markets requires our urgent and immediate attention...
>
> "We must not repose in fancied security that we can forever sell everything and buy little or nothing. If such a thing were possible, it would not be best for us or for those with whom we deal. We should take from our customers such of their products as we can use without harm to our industries and labor...

"The period of exclusiveness is past. The expansion of our trade and commerce is the pressing problem. Commercial wars are unprofitable. A policy of good will and friendly trade relations will prevent reprisals. Reciprocity treaties are in harmony with the spirit of the times, measures of retaliation are not."

President McKinley's speech signaled the beginning of the end of America's protective trade tariffs. For the next 125 years protective tariffs, except during three Republican presidencies—Warren G. Harding, Herbert Hoover, and Donald Trump—would steadily decline.

Largely forgotten today, was McKinley's role ending the bitter battle over gold and silver as the nation's monetary standard—specifically, whether the nation should base its currency on gold alone or allow both gold and silver. Long simmering, the battle exploded in 1873 when Congress passed the Coinage Act. The Act effectively ended the minting of silver dollars, placing the U.S. on a de facto gold standard.

Proponents of the gold standard, including most bankers, industrialists, and urban financial interests, argued that gold provided a stable and internationally respected currency. They believed that a gold-backed dollar would maintain the nation's credit and attract foreign investment.

On the other hand, many in the agrarian South and West, pushed for restoring silver as a second monetary standard. They advocated for the unlimited coinage of silver, at a ratio of 16 to 1 with gold, which would expand the money supply.

Why would expanding the money supply help workers and farmers? Because a growing money supply leads to inflation making it easier to repay the debts owed by workers and farmers—while hurting the bankers who hold the debt.

The conflict reached a climax in the 1896 presidential election. Democratic candidate William Jennings Bryan famously championed the silver cause with his "Cross of Gold" speech, condemning the gold standard for hurting the working man. But gold supporter William McKinley, financed by wealthy industrialists, won the presidency.

The debate ended with the Gold Standard Act of 1900 which formally committed the U.S. to gold and ended the push for silver as a monetary standard.

The battle over gold and silver had become such a major issue during the late nineteenth century that many historians believe it even found its way into a popular children's novel, *The Wizard of Oz*, published in 1900.

Prior to writing *The Wizard of Oz* the book's author, L. Frank Baum, had been a newspaper publisher and reporter with a flair for satire. Similarly, the book's illustrator, William Denslow, was a political cartoonist. The two men may have met at the Chicago Press Club where they were both members. Neither had ever written a children's book,

In one interpretation of the novel, *The Wizard of Oz* is a clever political allegory reflecting the debate over the gold and silver monetary standards that dominated American politics during the last decades of the nineteenth century.

Based on this interpretation, Dorothy represents the common American, kind, innocent, and unassuming. The cyclone represents the economic and political disruption during the eighteen-nineties. The Scarecrow stands for American farmers, portrayed as lacking brains but actually full of practical wisdom. The Tin Man represents industrial workers, dehumanized and rusted by hard labor and economic hardship. The Cowardly Lion stands for William Jennings Bryan, the populist politician who roared about free silver but lacked the political strength to win the presidency.

In the novel, Dorothy famously follows the yellow brick road to the Land of Oz to meet the Wizard. More than a happy stroll, the yellow brick road can be interpreted as the gold standard which leads to Oz, the abbreviation for an ounce (of gold).

The Emerald City reflects the illusion of prosperity under the gold standard. The Wizard represents America's politicians hiding behind a curtain of deceit.

Dorothy's silver shoes (changed to ruby red in the 1939 film) symbolize the power of silver to carry the people back home to prosperity.

Over the last 125 years, millions of children, and adults, have read *The Wizard of Oz* or seen the movie. It's a beloved American classic, and maybe a veiled history lesson.

L. Frank Baum never commented on his book's hidden meaning, if any. But more than a century later, it's fun to speculate that the political struggles during McKinley's presidency ironically inspired one of America's most beloved children's books.

Perhaps the bitter politics dividing America today will inspire someone to do the same.

# The Corpse at Every Funeral, the Bride at Every Wedding

APRIL 29, 2025

On April 22, the White House updated the covid.gov website. For years, the original federal website provided guidance on Covid testing, treatment, vaccines, and other information to counter the disinformation that flooded cable news and social media such as:

> "Now, I want to tell you the truth about the coronavirus... I'm dead right on this. The coronavirus is the common cold, folks."
> 
> Rush Limbaugh | February 2020

> "Between late December of 2020 and last month, a total of 3,362 people apparently died after getting the COVID vaccine in the United States. That's an average of roughly thirty people every day."
> 
> Tucker Carlson | May 2021

> "A global elite led by the CIA had been planning for years to use a pandemic to end democracy and impose totalitarian control on the entire world."
> 
> Robert F. Kennedy, Jr. | May 2022

"Moderna and Pfizer shots have a bovine protein in them that then creates an autoimmune response in those that take the shot so that many of them can no longer eat beef."

Alex Jones | August 2023

Starting on April 22, Americans accessing covid.gov were redirected to a White House page featuring a heroic image of President Trump before a bold headline, "Lab Leak: The True Origins of Covid-19."

The new web page—which redirects from the Department of Health and Human Services to the White House—is beautifully constructed and makes a convincing argument for the casual reader that the Covid virus had leaked from the Wuhan Institute of Virology.

The headline at the repurposed covid.gov website on April 22, 2025.
WHITE HOUSE

And above all, it features President Trump as a heroic warrior for truth. But can't the truth speak for itself? Apparently not, bringing to mind Alice Roosevelt Longworth's memorable description of her father, President Teddy Roosevelt: "My father always wanted to be the corpse at every funeral, the bride at every wedding, and the baby at every christening."

I wonder what Alice would have thought of Donald Trump?

The White House web page is based on the final report by the "Congressional Select Committee on the Coronavirus

Pandemic" released on December 4, 2024. Unlike the White House web page which confidently declares a lab leak as the source of the Covid virus, the Congressional report is more cautious, simply stating, "SARS-CoV-2, the Virus that Causes COVID-19, Likely [my emphasis] Emerged Because of a Laboratory or Research Related Accident."

After a two-year investigation, the evidence that the House Committee uncovered was exhaustive, but no "smoking gun" was found directly linking the Covid virus to a lab leak.

That shouldn't be surprising. It took fifteen years before virologists located the source of the virus responsible for the 2003 SARS outbreak—horseshoe bats dwelling in a remote cave located in China's Yunnan province. So, it may be years, if ever, before virologists conclusively identify the source of the Covid virus.

The Chinese, of course, disagree that the Covid virus leaked from the Wuhan Institute lab offering their own theory—that athletes from the U.S. military brought the virus into China. In a strange coincidence, the Seventh Military World Games were held in Wuhan, China during October 2019. The games attracted 9,300 military athletes from 109 countries. Two months later, the Covid virus was discovered in a Wuhan wet market selling fresh meat and live animals.

The Chinese have their own rap song describing the source of the virus, "Open the Door to Fort Detrick." Fort Detrick, located in Frederick, Maryland, has conducted research in bioweapons since World War II. The rap song's lyrics claim the Covid virus escaped from Fort Detrick:

> "Fort Detrick, more like a witch's cauldron. How many plots came out of your labs? How many dead bodies hanging a tag? What you're hiding, open the door to Fort Detrick. Cause transparency is your favorite. OK, great. America First. We want, want the truth."

No evidence though has been found that the military games were the source of the Covid virus.

If the Covid virus did not originate in the Wuhan Institute lab and first jumped to humans in the Hunan wet market, a troublesome question arises: How to account for the extraordinary coincidence that the virus happened to emerge from a wet market, one of thousands in China, located in the same city as the world's largest biological lab conducting "gain-of-function" research to make coronaviruses more infectious to humans?

In 2019, there were 39,400 wet markets located in China. If the Wuhan Institute's high-security laboratory had been conducting gain-of-function research on coronaviruses, the statistical probability that, out of the thousands of wet markets in China, the emergence of the Covid virus happened in the market nearest the Wuhan Institute of Virology is vanishingly small, equal to .0025 percent (1/39,400).

If the virus did emerge from the Wuhan wet market just a few miles from the Wuhan Institute of Virology, it was a remarkable coincidence.

In a serious omission, the White House website fails to mention that on December 19, 2017, the Trump administration lifted a three-year funding moratorium on gain-of-function research.

The Obama administration had instituted the moratorium on October 17, 2014. "In light of recent concerns regarding biosafety and biosecurity," the official statement declared, "effective immediately, the U.S. Government (USG) will pause new USG funding for gain-of-function research on influenza, MERS or SARS viruses." (MERS, SARS, and COVID are all coronaviruses.)

"Many scientists worried," the journal *Science* reported at the time, "that if [a] potent new lab strain were accidentally or deliberately released, it could result in a deadly pandemic…"

Three years later, on December 19, 2017, as part of the Trump administration's deregulation efforts, the National Institutes of Health (NIH) lifted Obama's funding moratorium. "[Gain-of-function] research," the NIH declared at the time, "is important in helping us identify, understand, and develop strategies and effective countermeasures against rapidly evolving pathogens that pose a threat to public health."

The 557-page Congressional report mentions that funding for gain-of-function research was restored in 2017—but only briefly on page 61 which referred to the Obama moratorium as the "2014 OSTP Pause."

According to the Congressional report, after lifting the moratorium, "The U.S. National Institutes of Health funded gain-of-function research at the Wuhan Institute of Virology." That's not surprising. Starting in 2014 the NIH had funded the Wuhan Institute to conduct research on China-based coronaviruses.

Today, we can only speculate. But future historians may consider the restoration of gain-of-function research on December 19, 2017, to be the singular event which led to the Covid pandemic and over seven million deaths worldwide.

# Measuring President Trump's First 100 Days in Office

MAY 6, 2025

Assessments of President Trump's first 100 days in office filled last week's news. Reports varied widely, of course.

To understand how widely, I collected the headlines and the first few paragraphs from eight leading news sources posted mid-day on April 29 and April 30. The sources range from the very conservative Breitbart to the highly liberal Huffpost with *The Wall Street Journal* and *The New York Times* anchoring the middle (at least in my opinion). For each news source, I selected the leading, top of the page, news article with the exception of *The Wall Street Journal* whose article was located in the opinion section.

This is a longer post than usual. So, if nothing else, scan through the headlines to see the steady progression from glowingly positive to darkly negative views of President Trump and his first 100 days in office.

How many of these news articles would pass Joseph Pulitzer's admonition?

> "What a newspaper needs in its news, in its headlines, and on its editorial page is terseness, humor, descriptive power, satire, originality, good literary style, clever condensation, and accuracy, accuracy, accuracy!"

SHORT ESSAYS FOR INQUIRING MINDS

# BREITBART

### Trump Touts 'Best' First 100 Days 'Of Any President in History'

by Nick Gilbertson | April 29, 2025

Trump has moved at a rapid pace through his first hundred days, signing at least 140 executive orders, including measures to secure the southern border, protect women's sports, and unleash America's energy potential, as a few examples.

Opinions regarding President Trump's first 100 days in office during his second term varied widely based on the news source.      BREITBART

The United States has also seen an influx of investment since his return to office, with the administration pinning the overall investment figure at over $5 trillion, which is estimated to produce over 450,000 jobs, while Trump and his deputies have worked to end two wars.

Trump called the last 100 days a "revolution in common sense" that transcends traditional political ideology.

"We're making America great again, and it's happening fast. What the world has witnessed in the past 14 weeks is a revolution of common sense; that's all it is," he said. "You're conservative, you're liberal, whatever the hell, you know what it's all about? It's about common sense, when you think about it."

"We like strong borders. We like good education. We like low interest rates. We like being able to buy a beautiful car and now deduct the interest on the loan," he added. "That's never happened before... We want a strong military. We want low taxes."

## NEW YORK POST

**Michael Goodwin: Trump's first 100 days illustrate his strengths and mindset —decisive, bold, and in a hurry**

by Michael Goodwin | April 29, 2025

Any discussion of the opening days of Donald Trump's presidency must start at the key date — last Nov. 5, when he rose from the political dead to seize his second term in the White House. His comeback victory was decisive as he swept all seven battleground states on the way to piling up 312 electoral votes, winning the popular vote and leading the GOP to control of Congress.

But first he had to survive two assassination attempts, with one in Pennsylvania a miraculous near miss, and overcome an onslaught of Democratic prosecutions and civil suits designed to defeat and imprison him.

All those cases, the first ever brought against a former president, were necessary, Americans were assured by Dems and their media mouthpieces, to protect democracy.

The Big Lie — that the weaponization of the courts was anything other than a partisan power play — seems like ages ago. But recognizing the dogged determination Trump needed to survive the persecution and come out on top is key to understanding his conduct since he took the oath on Jan. 20.

He believes God spared him to save America, and so his sense of mission is infused with urgency. He savors revenge — who wouldn't? — but ultimately came to get big things done.

He's in a hurry and sometimes, as with tariffs, to a fault.

## FOX NEWS

### White House lists dozens of 'hoaxes' pushed by media, critics in Trump's first 100 days

by Alexander Hall | April 29, 2025

The White House released a list of the "most egregious hoaxes" perpetuated by the media in the first 100 days of President Donald Trump's second term Tuesday.

The Trump administration published a press release declaring, "Since President Donald J. Trump took office 100 days ago, it has been a nonstop deluge of hoaxes and lies from Democrats and their allies in the Fake News suffering from terminal cases of Trump Derangement Syndrome."

The administration went on to list 57 purported "hoaxes" spread by the president's critics, the media and Democrats.

"HOAX: Rep. Eric Swalwell (D-CA) claimed 'no president' presided over more plane crashes during their first month in office as President Trump," was one example.

Data from the Department of Transportation indicated that more plane crashes occurred during the first few weeks of then-President Biden's term, as there were 55 aviation accidents in the U.S. between Biden's inauguration on Jan. 21, 2021, and Feb. 17, 2021, compared to 35 during the same period for Trump.

# THE WALL STREET JOURNAL

### At 100 Days, Trump 2.0 Is in Trouble

by The Editorial Board | April 29, 2025

Presidential second terms are rarely successful, and on the evidence of his first 100 days Donald Trump's won't be different. The President needs a major reset if he wants to rescue his final years from the economic and foreign-policy shocks he has unleashed.

There's no denying his energy or ambition. Mr. Trump is pressing ahead on multiple fronts, and he has had some success. His expansion of U.S. energy production is proceeding well and is much needed after the Biden war on fossil fuels. He has ended the border crisis in short order.

He is also rolling back federal assaults on mainstream American values—such as by policing racial favoritism. Mr. Trump was elected to counter the excesses of the left on climate, culture and censorship, and he is doing it.

On other priorities, the execution hasn't matched the promises. That would seem to apply to DOGE, which we've supported but has been so frenetic it isn't clear what it is achieving. Easy targets like USAID make for symbolic victories but no fundamental change in the growth of government. The Trump budget will offer more reform proposals, if the White House can get them through Congress. He badly needs a pro-growth tax bill.

# THE NEW YORK TIMES

### Trump's Presidency So Far

By Irineo Cabreros, Aatish Bhatia | April 30, 2025

The first 100 days of Donald Trump's second presidency have been a study of extremes, especially when compared with the start of presidential terms over the last century. Today, The Upshot — a section of The Times focused on data and policy — published eight charts comparing Trump's performance with that of his predecessors...

On his first day in office, Trump signed a record 26 executive orders — and he didn't stop there. The executive order has become something of a hallmark of his governing style, a way to express clear policy directives without the bureaucracy of regulation or the horse trading of legislation.

Some orders direct federal agencies to develop policy in particular areas, like oil drilling, prescription drug prices or the water pressure delivered by shower heads. Some mostly express the president's sentiment on an issue. Some function as warnings or punishments for political enemies. But many — in key areas like immigration and tariffs — effectively carry the force of law. Compare the president's output with that of Congress, which has passed only a handful of laws since Trump's inauguration.

Trump's executive actions have already led to an explosion of lawsuits. In other recent administrations, the suits have come later, in response to laws and regulations that take months and years to develop. But Trump is moving quickly to cut funding, fire federal workers, impose tariffs, reshape immigration policy and more.

## CNN

### Historic and controversial changes at breakneck speed: Inside Trump's first 100 days

by Alayna Treene, Kevin Liptak | April 29, 2025

Delivering the longest inaugural address in history in January, newly inaugurated President Donald Trump made clear he had little time to waste.

"From this moment on, America's decline is over," he said, before adding: "All of this will change starting today, and it will change very quickly."

One hundred days later, Trump has found mixed success fulfilling the pledges in his speech to return "faith, wealth, democracy and freedom" to a beleaguered nation. Americans have grown increasingly skeptical, and his 41% approval rating in CNN's latest poll is the worst for any president at his 100-day mark – including himself, in 2017.

Yet few would argue he hasn't met his promise of speed.

Despite only signing one piece of legislation in a ceremony at the White House, Trump has ushered in the most dramatic change of any president in decades, transforming the nation's economy, foreign policy, federal workforce and immigration enforcement in ways that left his opponents gasping. Working at breakneck pace and awake to lessons from his first term, he has pursued almost all of his agenda through executive actions.

## MSNBC

**There's a recipe for a successful first 100 days. Trump is missing an ingredient.**

by Hayes Brown | April 29, 2025

By the standards of his predecessors, President Donald Trump's first 100 days back in office falls far short in terms of lasting achievements — and his overall popularity, for that matter. The phenomenon of the first 100 days in office became political legend, and hence, a milestone marker for the modern presidency, after President Franklin D. Roosevelt's explosive first term. But this somewhat arbitrary measurement has helped foster a profound misunderstanding of the way the federal government was designed to work.

The mythic air surrounding Roosevelt's first 100 days established it as a standard for executive prowess, given the speed at which his New Deal program coalesced and helped stem the worst effects of the Great Depression. But the plaudits the White House received for that burst of energy should be at the very least shared with Congress.It was on Capitol Hill that the New Deal went from concept to law, a flexing of legislative muscle and cornerstone for Roosevelt's success. As we look at Trump's decidedly unimpressive start, it's worth remembering that without a willing and ready Congress to act in concert with FDR, the New Deal would have been no deal at all.

When Roosevelt came into office at the start of 1933, the country was no longer in economic free fall but saw little stability or sign of recovery. The ongoing ripples of the stock market crash almost four years prior still left tens of millions of Americans unemployed and banks teetering on the verge of collapse. President Herbert Hoover took much of the political blame for his sclerotic response. His fellow Republican lawmakers also paid the price at the ballot box the previous fall, having passed harsh tariffs that accelerated the economic devastation.

## HUFFPOST

### Trump's First 100 Days: Destroying America Was The Plan All Along

by Paul Blumenthal | April 30, 2025

In his first 100 days, President Donald Trump has taken a bludgeon to the government, civil society, civil rights laws, efforts at desegregation, foreign alliances, anti-corruption law and norms, the entire global economy, and the country's very self-conception as born from the idea that "all men are created equal."

But for all the questioning of whether Trump knows what he is doing, this is not an accident, or a byproduct of other goals: Destroying America as we know it was always the point of the second Trump administration. That country was born in the 20th century, and Trump's second term agenda, led by a group of the ideologues and oligarchs who backed his reelection, aims to repeal the 20th century.

The 20th century brought the United States the greatest wealth and success it had ever seen and to its pinnacle of global influence. Following World War II, the country created the best university system in the world which worked in tandem with the government on scientific research and development, positioning it as an academic and intellectual powerhouse. The country also saw a rights revolution, with legislators and judges extending the promise of the Declaration of Independence to more people than ever before and cementing the U.S.' status as the moral leader of the "free world."

This refounding of the country, that ran from Franklin Roosevelt's New Deal through the Civil Rights Movement, made the country richer, more equal, more educated and more successful than ever before. But not everyone accepted progress at the time, and they pushed to turn back the clock.

---

In 1921, Albert Einstein visited the United States. An acute social observer, Einstein, describing his first impressions of America, wrote:

> "The [American] Press, which is mostly controlled by vested interests, has an excessive influence on public opinion."

> Albert Einstein | *Some Notes on My American Impressions*, 1921

# President Trump and Pope Leo XIV

MAY 13, 2025

Last Thursday, May 8, had personal significance for me; Cardinal Robert Prevost was elected Pope Leo XIV.

Although a confirmed Lutheran, I graduated from a Catholic high school, Cascia Hall Preparatory School located in Tulsa, Oklahoma. While there, I inevitably absorbed some Catholic theology. So I found it interesting that from 1999 through 2001, Father "Bob" Prevost served on Cascia Hall's board of directors. That's not surprising as both Cascia Hall and Robert Prevost are Augustinians, a Catholic order founded in 1244 with a mission to promote social justice and service to the poor—a mission that often brings Augustinians into conflict with authoritarians.

Early in his priesthood, Father Prevost spent twenty years as a missionary in Peru where he conflicted with Peru's president, Alberto Fujimori. Fujimori was an authoritarian who in 1992 dissolved Peru's Congress, suspended its constitution, and ordered the extrajudicial killings of his opponents. In an effort to promote (white) racial purity, Fujimori ordered the forced sterilization of as many as 300,000 indigenous women. Prevost bravely condemned Fujimori's authoritarian government, its killings, and forced sterilizations.

Robert Prevost's condemnation of authoritarianism in Peru suggests as pope he will advocate for those who cannot speak out for themselves.

That was certainly the case with his predecessor, Pope Francis, who criticized Vice-President Vance, a recent convert to Catholicism, for his misinterpretation of *Ordo Amoris* (Latin for "order of love"). Vance had invoked the medieval Catholic theology to justify President Trump's immigration and foreign aid policies. "We should love our family first," Vance declared during a Fox News interview with Sean Hannity, "then our neighbors, then love our community, then our country, and only then consider the interests of the rest of the world."

Robert Prevost spent twenty years in Peru as a missionary.
VATICAN NEWS

Pope Francis disagreed with Vice-President Vance's interpretation of *Ordo Amoris*, and in a letter to the U.S. bishops wrote:

> "Christian love is not a concentric expansion of interests that little by little extend to other persons and groups... The true *ordo amoris* that must be promoted is that which we discover by meditating constantly on the parable of the 'Good Samaritan', that is, by meditating on the love that builds a fraternity open to all, without exception."

Most Christians know this principle simply as, "Thou shalt love thy neighbour as thyself." (Matthew 22:39)

Early indications are that Pope Leo XIV may follow a similar path as Pope Francis in condemning President Trump's often cruel deportation policies. Prior to his election as pope, Robert Prevost had been active on social media. On April 14, for example, Prevost reposted a comment critical of President Trump and Salvadoran President Nayib Bukele for joking about the wrongful deportation and detention of Kilmar Abrego Garcia. The tweet asked, "Do you not see the suffering? Is your conscience not disturbed? How can you stay quiet?"

In his first formal meeting with the cardinals, Pope Leo XIV said he chose his papal name to continue the social activism of Pope Leo XIII whose papacy extended from 1878 through 1903, a period of rapid industrialization.

It was also a period of great social unrest. In 1877, twenty men who belonged to the secret labor organization of Irish miners known as the "Molly Maguires" were hung. The men had been convicted in a rigged trial for alleged murders committed during violent labor strikes after their Pennsylvania mining companies cut wages. That same year, ten striking workers died in a nationwide railroad strike.

A decade later, in 1886, seven policemen and four striking workers were killed in a Chicago strike over wages and working conditions. Then, in 1892, twelve men died during the infamous Homestead Steel Works Strike.

It was during this period of tumultuous labor unrest that Pope Leo XIII issued *Rerum Novarum* ("Of New Things") on May 15, 1891, his monumental encyclical which discussed the rights and duties of capital and labor.

Leo XIII began the encyclical by acknowledging the conflict between capital and labor, criticizing both unregulated capitalism and socialism. He defended the right to private property while emphasizing that wealth must be used responsibly and not at the expense of workers. The encyclical upheld the dignity of work and insisted on just wages, rest, and humane working conditions. It affirmed the right of workers to form unions to protect their interests. Leo XIII stressed the importance of cooperation between classes and the role of the family as the cornerstone of society. He warned against class hatred and encouraged mutual respect between employers and employees. *Rerum Novarum* ultimately promoted a vision of society rooted in Christian values—justice, charity, and the common good.

Although it would take decades for labor and capital to come to an uneasy peace, Pope Leo's *Rerum Novarum* laid the groundwork for the Church's future role advocating for social justice and human dignity.

Shortly after the announcement of Robert Prevost as the first American pope, President Trump issued a statement on Truth Social declaring:

> "Congratulations to Cardinal Robert Francis Prevost, who was just named Pope. It is such an honor to realize that he is the first American Pope. What excitement, and what a Great Honor for our Country. I look forward to meeting Pope Leo XIV. It will be a very meaningful moment!"

Meaningful indeed. It's not possible to imagine two more different men. One whose life has been guided by humility, compassion, and theological doctrine. The other the accumulation of fame, fortune, and political power.

As the first American pope, Pope Leo VIV has received a harsh reception from MAGA supporters. Laura Loomer, a far-right provocateur, called the new pope, "anti-Trump, anti-Maga, pro-open Borders, and a total Marxist like Pope Francis." Steve Bannon, one of the President's strongest supporters, called Pope Leo XIV "the worst pick for MAGA Catholics" and "an anti-Trump by the globalists who run the Curia—this is the pope [Pope Francis] and his clique wanted." Bannon predicted tensions between the White House and Vatican that could even tear apart American Catholics.

Today, we can't know how a conflict between President Trump and Pope Leo XIV might resolve itself. Let's hope that two of the world's most powerful men, one politically, one religiously, find common ground.

# On Workers and Capital

MAY 20, 2025

I had an interesting Uber ride last week. It was a short, twenty-minute ride from the center of Washington, D.C. to Reagan Airport. As usual, I enjoyed chatting with the driver, a hard-working immigrant with an Arabic first name and German last name whom I'll call "MK."

Weird as it sounds, my chat with MK brought to mind Pope Leo XIII's encyclical regarding the relationship between workers and capital discussed in my recent Substack. Uber, it turns out, is an extreme example of the division that often exists between workers and the capital that finances them.

MK is a full-time Uber driver and likes his job, especially the freedom to set his own hours with no boss looking over his shoulder. Well, that's not quite true. The Uber driver app tracks every driver action including acceleration, braking, cornering, and speed. And, who knows, maybe even conversations? Over the years, multiple Uber drivers, and their passengers, have been charged with sexual harassment.

MK told me he tries to earn two-hundred dollars a day after Uber takes out their share. Assuming he works five days a week, and fifty weeks a year, MK's annual gross income is roughly $50,000. That's before his auto, gas, insurance and other expenses which I'll estimate at $7,000. That puts MK's annual take-home pay around $43,000, comfortably above the $32,500 poverty level for a family of four, and well above $15,060 for a single person.

So, MK is doing nicely navigating today's gig economy. He is fortunate though compared to most Uber drivers. MK works full-time in the nation's capital, a beehive of politicians, lobbyists, federal employees, and tourists.

MK and his fellow drivers represent one side of the economic equation: the workers. What about the entrepreneurs and early investors who put up the "sweat equity" and money that made Uber possible?

Uber CEO Travis Kalanick (second from left) and the Uber team circa 2011.                                                                                              UBER

It all started one rainy night in Paris when Travis Kalanick and Garrett Camp were unable to find a taxi after attending a business conference. That led to a seemingly simple idea: connect riders and drivers with a phone app. Rather than standing in the street trying to hail a taxi, just summon one on your smart phone.

Little more than a year later, Kalanick and Camp had founded Uber, raised a $5.4 million seed investment, and would soon launch their "UberCab" ride-hailing service in San Francisco.

With the taxi industry deeply entrenched, it was long, bitter battle city by city to overcome local opposition. But many of Uber's early challenges were self-inflicted. Uber's testosterone-driven management was accused of an aggressive, "win at all costs" business philosophy, a toxic workplace culture, and widespread sexual harassment.

An early photo (above) of a young woman wedged behind seven men is suggestive of Uber's early male-dominated culture.

By 2017, the public scandals had become so damaging that Travis Kalanick was forced out of the company. He was replaced by Dara Khosrowshahi, former CEO of Expedia, to lead a cultural reset and prepare the company for a public offering.

Two years later, on May 10, 2019, Uber went public. Travis Kalanick and Garrett Camp collectively made $9.0 billion. Uber's earliest investors also did fabulously well making a fortune many times over—their $5.4 million seed investment in 2010 was valued at $7.4 billion the day Uber went public. In just one example, Mitch Kapor (who developed Lotus 123, the first killer app on the IBM PC) invested $95,000. Nine years later, Kapor's investment was worth $372.4 million.

A May 10, 2019, *Wall Street Journal* article called Uber's $5.4 million seed investment "one of the greatest venture-capital investments of all-time—even when compared with the famous early bets on Google, Facebook and startups from the dot-com boom."

Today, after revolutionizing the moribund taxi industry, Uber has a grand vision to become "the indispensable digital infrastructure for how people and goods move in cities—faster, cleaner, and smarter."

As a former technology entrepreneur, I'm supportive of the huge financial gains Uber's founders, early employees, and initial investors enjoyed. I know from experience that Uber's early employees for years worked long hours at salaries well below market. Not to be forgotten, the company's early investors took an astronomical risk betting that Uber's young founders would revolutionize urban transportation.

Fortunately, it all paid off.

But what about more traditional companies like, say, Walmart? What's their responsibility to workers and capital?

As I've written before, starting around 1970, American companies began following a business philosophy promoted by the economist Milton Friedman. Friedman declared that corporations have no social responsibility beyond increasing profits. Friedman provided corporations a simple, unequivocal yardstick for measuring success:

> "There is one and only one social responsibility of business, to use its resources and engage in activities designed to increase its profits so long as it stays within the rules of the game, which is to say, engages in open and free competition without deception or fraud."

By the late nineteen-eighties, Friedman's philosophy was well entrenched in corporate boardrooms. The result, as depicted in the above chart, was that those able to generate financial gains—entrepreneurs, management, investors—enjoyed rapidly rising incomes while others, the nation's workers, languished.

A 2013 Congressional report singled out Walmart for paying sales associates an average of $8.81 an hour, wages that placed their workers near poverty levels allowing them to qualify for supplemental income and food benefits. These people weren't deadbeats feeding off government largess, but workers with full-time jobs with one of America's largest corporations. The report estimated that federal and local governments paid an average of $904,542 per year in welfare benefits to Walmart employees for each Walmart Superstore, or over $3,000 per employee.

That same year, Walmart's CEO, Mike Duke, received compensation amounting to $20.7 million. That's a lot of money. But Walmart reported $469 billion in revenue and $17 billion in net income that year. So, in my opinion, Duke was not overpaid earning just 0.1 percent of Walmart's net earnings.

Walmart's payment polices began to change in 2015 when the company announced that it would raise wages for entry-level employees to nine dollars an hour and then to ten dollars by early 2016. Walmart had come to realize that better pay was good business. "Customers need to be served," Walmart's CEO, Doug McMillan said at the time, "and associates need to be happy and love their job." To their credit, Walmart executives exceeded their promise when the company raised the hourly wage for 1.2 million Walmart workers to $13.38 in January 2016.

Today, according to Walmart, "the average hourly wage for our U.S. frontline associates is close to $18.00." That's below the poverty level for a family of four, but more than sufficient for a frugal individual.

The issue of the appropriate worker compensation has existed since before the dawn of the industrial age. Should workers be considered a commodity whose compensation is determined solely by the market as Milton Friedman argued? Or do employers have a responsibility to provide their workers a minimum living wage as Pope Leo III argued in his 1891 encyclical, *Rerum Novarum*:

> "Let the working man and the employer make free agreements, and in particular let them agree freely as to the wages; nevertheless, there underlies a dictate of natural justice more imperious and ancient than any bargain between man and man, namely, that wages ought not to be insufficient to support a frugal and well-behaved wage-earner."

# How to Make America Great Again

MAY 27, 2025

Nobody seems to know what period President Trump has in mind when he talks about Making America Great Again.

I'm guessing it's the nineteen-fifties. All the other decades were either too recent or marred by war, scandal, or economic crises. But the fifties, the fifties are remembered as a period of peace and prosperity. Traditional American families laughed at *I Love Lucy*, admired John Wayne's swagger, saw polio conquered, watched Dinah Shore sing *See the USA in your Chevrolet*, and danced to rock-n-roll.

Most Americans had good-paying jobs. Jobs that could comfortably support a family of four with a new family car every four or five years, a modest family vacation in the summer, and plenty of presents under the Christmas tree. If you could manage to forget the threat of nuclear war, life was secure.

That is, if you were White and Protestant. The country was deeply divided racially with negroes, as Blacks were then known, living, attending school, and socializing in segregated communities. No Catholic had yet served as president. Jews were limited, or banned, at many universities, hospitals, and law firms.

Small towns in America had well-defined social strata. The owner of the local factory drove a Cadillac, the bank president a

Buick, doctors and lawyers an Oldsmobile, and most everybody else Fords and Chevrolets. A few iconoclasts drove Plymouths.

Personal incomes were relatively well distributed. In 1956, the average after-tax income (federal taxes only) for the middle 80 percent of taxpayers was $25,500 (2020 dollars) while the top 1.0 percent took home $213,000, only about eight times more. Few Americans resented the difference; the wealthy in town generally ran the businesses that provided jobs for everybody else.

A typical 1950s family enlivened by a mischievous young daughter.

Today, America is far different. In 2022, the middle 80 percent of taxpayers took home about $45,500 while those in the top 1.0 percent bracket averaged $1.32 million, nearly thirty times more. That's the largest income gap since the late nineteen-twenties shortly before the 1929 Stock Market Crash and the ensuing Great Depression.

During the nineteen-fifties, Americans trusted their government. After all, the nation's leaders had led the country out of the Great Depression and won the Second World War. America was both loved and feared but, above all, respected around the world.

That respect extended to President Eisenhower. Eisenhower had been the Supreme Commander during the Second World War leading the Allied forces to victory against Germany. In 1952 during his presidential campaign, Eisenhower promised to end the Korean War. Six months after taking office, the Korean Armistice Agreement was signed on July 27, 1953, initiating an uneasy peace that has lasted for over seventy years.

Eisenhower had the integrity to take unpopular positions. Few issues stir more emotion than taxes. That hardly deterred Eisenhower. A month after assuming the presidency, President Eisenhower told Americans he would not lower taxes until the federal budget was balanced:

> "There must be balanced budgets before we are again on a safe and sound system in our economy. That means, to my mind, that we cannot afford to reduce taxes, reduce income, until we have in sight a program of expenditures that shows that the factors of income and of outgo will be balanced. Now that is just to my mind sheer necessity.... I have as much reason as anyone else to deplore high taxes.... But I merely want to point out that unless we go at it in the proper sequence, I do not believe that taxes will be lowered."

Eisenhower's position on taxes was not unique. From the end of the Second World War through the late nineteen-seventies, managing the federal debt was an important element of federal tax policy. In 1946, a year after the end of the Second World War, the federal debt stood at 118 percent of the nation's

Gross Domestic Product (GDP). Over the next thirty-five years, sound fiscal management by the federal government reduced the debt to 32 percent of GDP. Had that trend continued, the United States would have been debt-free by the year 2000.

But starting around 1980, that trend reversed. Spending, after averaging 18 percent of GDP from 1946 through 1980, began to climb. Over the next forty-five years, federal spending averaged 21 percent of GDP hitting a high of 31 percent during the Covid pandemic.

Yet, federal taxes failed to keep up with the additional spending. Rather than rising, tax rates actually fell starting with the Reagan tax cuts in 1981.

That was especially true for corporate income taxes which averaged 3.1 percent of GDP from 1946 through 1980. But, after steep tax cuts, were reduced to an average of 1.6 percent of GDP from 1980 through 2024.

The reduction in corporate income taxes cost trillions in federal tax revenues. Had tax rates been maintained at their 1946-1980 average level, an additional eleven trillion dollars in tax revenues would have been collected in the decades after 1980—$567 billion in 2024 alone.

But aren't tax cuts supposed to increase tax revenues as Treasury Secretary Andrew Mellon a century ago and economist Arthur Laffer more recently have claimed?

That's true, but only when tax rates are too high. High tax rates discourage investment and work since much of the gains are taxed away. Hence economic growth slows and tax revenue declines.

But moderate tax rates have little effect on economic behavior. Taxes are just another cost of doing business. So, cutting tax rates has little effect on economic growth and just reduces tax revenues. That apparently was the case when corporate taxes were consistently reduced starting around

1980. Lower taxes failed to boost economic growth and simply reduced tax revenues.

Skeptical? From 1946 through 1979 real GDP, adjusted for inflation, grew at an average rate of 3.9 percent annually. Corporate taxes averaged 3.1 percent of GDP. But after 1980, annual GDP growth slowed to 2.6 percent and corporate taxes fell to 1.6 percent of GDP.

That's the big picture. But even a brief analysis of President Trump's 2017 tax cuts confirms the loss in tax revenue overly large tax cuts have on tax revenues. That year the maximum corporate tax rate was reduced from 39 to 21 percent. During the next five years from 2017 through 2021 corporate taxes fell to an average 1.1 percent of GDP compared to 1.8 percent of GDP during the five years preceding the tax cuts.

As Congress passed tax cuts, many American companies chose not to invest their tax savings in their businesses building factories and hiring workers—business investment has been largely unchanged since 1950 averaging about 17 percent of GDP.

Instead, companies have channeled their tax savings to their (largely affluent) shareholders through share buy-backs and dividends. These windfall shareholder payments, paid for by tax cuts, significantly contributed to the large increase in wealth among affluent Americans since 1980.

But we all pay a price for a federal government that seems to be addicted to tax cuts while incurring huge budget deficits. In 2024, interest on the federal debt was $880 billion—over $2,500 for every man, woman, and child in America.

That's a huge expense every American is paying, if indirectly. Even more serious, today the national debt exceeds our annual GDP and continues to grow, all financed by the U.S. Treasury selling Treasury debt.

If the current trend continues, it may become impossible to sell that debt at anything less than exorbitant interest rates.

When that happens, the United States will be faced with an unprecedented financial crisis with unpredictable consequences.

Since we Americans have tolerated large budget deficits for decades, that reckoning will likely come as a surprise. As Ernest Hemingway wrote in his 1926 novel, *The Sun Also Rises*:

> "How did you go bankrupt? ...Two ways: Gradually, then suddenly."

# It's Déjà Vu All Over Again

JUNE 3, 2025

America's national debt is like the weather. Everybody talks about it but nobody does anything about it.

Well, that's not quite true. In 1993, newly-elected President Clinton made repairing the economy his top priority. The national debt, measured as a percentage of the Gross Domestic Product (GDP), had doubled during the twelve years preceding Clinton's presidency, from 32 percent of GDP in 1980 to 62 percent in 1992. It was the steepest rise in the national debt since the Great Depression.

Seven months after taking office, Clinton signed the Omnibus Budget Reconciliation Act which raised the maximum individual tax rate from 31 to 39.6 percent, corporate taxes from 34 to 35 percent, the federal gas tax from 14.1 to 18.4 cents per gallon, as well as other changes to the tax code.

Clinton's political opponents predicted the tax increases would quickly wreck the economy. "We have all too many people..." House Minority Whip Newt Gingrich claimed, "who are talking about bigger government, bigger bureaucracy, more programs, and higher taxes. I believe that will in fact kill the current recovery and put us back in a recession."

The Heritage Foundation was equally pessimistic. "If enacted," a Heritage report predicted on February 18, 1993, "the Clinton tax hike will fuel more federal spending, destroy jobs, undermine America's international competitiveness, reduce economic growth, and increase the budget deficit." (Today, the

Heritage Foundation is the author of Project 2025 steering President Trump's political agenda.)

The skeptics were wrong. Rather than withering, the economy blossomed. From 1993 through 2000, real GDP growth averaged 3.9 percent annually, the largest increase since the nineteen-sixties. Measured as a percentage of GDP, the national debt declined from 64 percent to 55 percent—a remarkable turn-around after increasing from 32 percent to 62 percent during the preceding Reagan and Bush administrations.

Although they had a rocky relationship, House Speaker Gingrich and President Clinton worked together to balance the federal budget in 1998.
GETTY

Clinton's economic policies were critical, but Clinton was also lucky. The Cold War had ended, allowing reductions in defense spending. Baby Boomers, in their peak earning years, were generating large Social Security tax surpluses. The internet-based technology boom drove employment, corporate profits, and the stock market to record highs.

Still without President Clinton's partnership with Congress, the necessary tax and spending reforms would never have been enacted. The combination of increased tax revenues and lower spending (supported by Republicans' "Contract with America" legislative agenda) resulted in President Clinton's signature economic achievement, a $559 billion budget surplus from 1998 through 2001. Other Presidents had grown the economy, cut taxes or slashed inflation. But no president since Calvin Coolidge back in the Roaring Twenties had managed to string together four consecutive budget surpluses as Clinton did from 1998 through 2001.

Today, the United States is confronted with a financial crisis far worse than President Clinton inherited. The national debt stands at 122 percent of GDP, a level even exceeding its peak during the Second World War. Interest payments on the debt exceed three percent of GDP—about $12,000 annually for a family of four—and are rising rapidly. Worst of all, America is as deeply divided as any time since the nineteen-sixties when racial integration, the Vietnam War, and cultural issues tore the nation apart.

Yet, Congress and the President need to come together to manage the federal deficit before a crisis erupts far worse than the 2008 financial crisis. Difficult, unpopular decisions to cut spending, and likely raise taxes, must be made to close the budget gap.

A good place to start would be the federal budget during 1998 through 2001 when the federal budget was last balanced. As the table below shows, federal tax revenues as a percentage of GDP today have declined sharply while federal expenditures have ballooned.

From 1998 through 2001, federal tax revenues averaged 19.1 percent of GDP while expenditures averaged 17.7 percent. Revenues comfortably exceeded expenditures and the nation enjoyed rare federal surpluses.

Today, the situation is reversed. From 2018 through 2024—excluding the pandemic years 2000, 2021, and 2022—federal tax revenues averaged 16.4 percent of GDP, while expenditures averaged 21.8 percent. Expenditures significantly exceeded revenues adding $5.3 trillion to the national debt in these four years alone.

That increase was predictable. Expenditures have increased since the pandemic from an average 20.5 percent of GDP in the five years preceding the Covid pandemic to over 23 percent in 2024. Most of the increase is due to rising costs in mandatory programs such as Social Security, Medicare, and Medicaid, as well as growing interest payments on the national debt.

Average Federal Taxes and Expenditures as a Percentage of GDP

|  | 1998 - 2001 | 2018 - 2024* |
|---|---|---|
| **Federal Taxes** | | |
| Personal Income | 9.4% | 8.1% |
| Corporate Income | 1.9% | 1.4% |
| Social Security | 6.4% | 5.8% |
| All Other | 1.5% | 1.1% |
| Total | 19.1% | 16.4% |
| **Expenditures** | | |
| Defense | 2.9% | 3.1% |
| Social Security | 4.1% | 4.9% |
| Medicaid | 2.0% | 3.0% |
| Medicare | 1.5% | 2.9% |
| Interest | 2.3% | 2.3% |
| All Other | 4.9% | 5.7% |
| Total | 17.7% | 21.8% |
| **Federal Surplus** | 1.4% | -5.4% |

* Excluding pandemic years 2020 - 2022

U.S. DEPARTMENT OF THE TREASURY

Either expenditures need to be reduced, or taxes need to be increased. Actually, both need to be done.

Although raising taxes is difficult—and hasn't been done since the Clinton administration—restoring taxes to their average rates during 1998 through 2001 would have increased tax revenues in 2024 by $504 billion reducing the budget deficit from $1.83 billion to $1.3 billion.

Reducing expenditures to their 1998 through 2001 levels would save far more, approximately $1.66 trillion. But that is unrealistic since America's population has aged and Social Security, Medicare, and Medicaid enrollment has swollen. Social Security enrollment alone has increased from 45.4 Americans in 2000 to 68.5 million in 2024.

Still, reducing expenditures is a huge opportunity. Just reducing expenditures to their 2015 - 2019 average percentage of GDP would have reduced federal expenditures in 2024 by a trillion dollars. That, coupled with restoring tax rates to their 1998 - 2001 levels, would have slashed the 2024 federal deficit from $1.3 trillion to $300 billion.

President Clinton's balanced budget was short-lived. In 2001, the internet bubble burst. Far worse, America on September 11 suffered the worst foreign attack on American soil since the Japanese bombed Pearl Harbor on December 7, 1941. To stimulate the economy, President George W. Bush cut taxes, slashed regulations, and increased government spending. It didn't work. By 2009 as Bush was leaving the presidency, the economy had crashed, and the national debt had soared to 82 percent of the GDP.

Today, President Trump is following a similar path as President Bush. As the nation continues to recover from the Covid pandemic, President Trump has promised to cut taxes and slash regulations. But the President also promised he "will not touch" Social Security, Medicare, and Medicaid—together, half of all federal expenditures.

As the great baseball philosopher, Yogi Berra, famously said, "It's déjà vu all over again."

# The Surprising Origins of Social Security

JUNE 10, 2025

On January 31, 1940, the first Social Security benefit, $22.54, was paid to Ida May Fuller of Ludlow, Vermont. Over the next thirty-five years, Ida collected a total of $22,888.92 in benefits until her death in 1975 at the age of 100.

In the years since Ida Fuller received her first check, Social Security has grown to become the federal government's largest, and most expensive, program. In 2023, Social Security paid $1.33 trillion in benefits to 71.6 million Americans.

Social Security marked the first time the federal government took long-term responsibility for the economic security of individual citizens, particularly the elderly, unemployed, and disabled. As part of President Franklin D. Roosevelt's New Deal, many historians consider Social Security to be one of Roosevelt's greatest legislative accomplishments—if not the greatest.

But a huge, mandatory government program also bred fierce critics.

Then:

> "Never in the history of the world has any measure been brought in here so insidiously designed so as to prevent business recovery, to enslave workers, and to prevent any possibility of the employers providing work for the people."
>
> —Representative John Taber (R-NY), April 19, 1935

And now:

"I think the government is one big Ponzi scheme if you ask me [and] Social Security is the biggest Ponzi scheme of all…"
— Elon Musk, March 24, 2025

General Electric's respected president, Gerard Swope, played a pivotal role in the development of Social Security.   WIKIMEDIA

There is a kernel of truth in both of these criticisms. From its inception, critics have considered Social Security to be governmental overreach bypassing the free market and depriving the private sector of capital and labor. And resonant of a Ponzi scheme, Social Security is a "pay as you go" program; benefits paid to retirees are funded by taxes paid by current workers rather than from retirement funds collected in advance.

But to understand Social Security, we need to appreciate, and respect, the tremendous hardships much of America, both the rich and the poor, faced during the Great Depression.

From the stock market crash in October 1929 through July 1932, the Dow Jones average fell from 381 to 41. Imagine today's market falling from a high of 45,000 in December 2024 to 4,866 in May 2027. As the financial markets collapsed, employment fell from 35.7 million workers in 1929 to 27.9 million in 1932 throwing one in four Americans out of work. It would take over a decade, and the Second World War, for employment to return to its 1929 level.

Millions of Americans had lost their jobs, their homes, and hope. Even former President Calvin Coolidge whose growth-oriented policies had ushered in the Roaring Twenties had lost hope. "In other periods of depression," Coolidge wrote in 1933, "it has always been possible to see some things which were solid and upon which you could base hope, but as I look about, I now see nothing to give ground to hope."

Years later in an October 13, 1982, speech, President Reagan recalled his experiences during the Great Depression:

> "I was 21 and looking for work in 1932, one of the worst years of the Great Depression. And I can remember one bleak night in the thirties when my father learned on Christmas Eve that he'd lost his job. To be young in my generation was to feel that your future had been mortgaged out from under you..."

President Carter's childhood during the Great Depression was equally bleak:

> "In the Great Depression in which I grew up and remember vividly, unemployment was over 25 percent, and over 35 percent where I lived. A grown man would work all day, 16 hours, for a dollar. I remember hundreds of people walking by, people who had come down from the North just

to get warm. They would come to our house as beggars even though they might have a college education."

Had capitalism failed? After losing hope during the depression, many Americans believed it had. As Americans lost confidence in capitalism they turned to the government for help.

In 1933, Dr. Francis Townsend, a retired physician, was one of the first to propose a federally funded, national pension system. Townsend proposed that all individuals over sixty who were retired and "free from habitual criminality" be paid $200 a month. Townsend claimed his plan would encourage older workers to retire and open positions for unemployed younger workers. By 1935 there were 7,000 Townsend Clubs with over 2.2 million members promoting the plan.

In 1934, the author Upton Sinclair ran for governor of California under his "End Poverty in California" (EPIC) campaign. Sinclair proposed California take over closed factories, abandoned farms, even movie studios and operate them to maximize employment rather than profit. The EPIC program was wildly popular with 800 EPIC clubs springing up in California within a year.

An even more extreme reformer was Huey Long, Louisiana's populist governor and senator, known as "the Kingfish." Huey Long's "Share Our Wealth" proposal called for the government to guarantee every family a generous annual income of $5,000 ($115,000 in 2025 dollars). Huey Long proposed paying for his program by distributing wealth from the rich to the poor, limiting annual incomes to $1.0 million ($23.3 million today), inheritances to $5.0 million and family fortunes to $50 million. By 1935, there were more than 27,000 Share Our Wealth clubs in the country with over 7.5 million members.

But the most consequential proposal was made by one of the nation's most respected business leaders, Gerard Swope. Swope had been the president of General Electric since 1922. He was also the chairman of the influential Business Advisory Council (which today endures as the Business Council comprising America's top CEOs).

In 1931, Swope proposed a national recovery plan to help pull the nation out of the depression. The plan proposed that all companies with over fifty employees and engaged in interstate commerce provide employees a common set of benefits including disability, unemployment, and life insurance plus a pension plan for retirement at age sixty-five. The company plans would be administered by trade organizations; Swope knew workers no longer trusted their employers to protect their pension savings.

The nation's leading business association, the Chamber of Commerce, supported Swope's plan as did other business organizations. But President Hoover did not, calling it compulsory, inefficient, and monopolistic.

Lacking President Hoover's support, the Swope plan went nowhere for three years. Then, on March 8, 1934, President Roosevelt and Gerard Swope had lunch at the White House. During the meeting, Swope advocated for a modified version of his 1931 plan. Roosevelt was intrigued and asked Swope to prepare a memo outlining his plan.

Two weeks later Swope delivered a memo that included a detailed statistical analysis for unemployment, disability, and old-age pension programs. Roosevelt's advisors felt aspects of Swope's plan were too ambitious. But Swope's proposal, both as General Electric's president and chairman of the Business Advisory Council, gave Roosevelt confidence to proceed with a less ambitious plan.

Three months after meeting with Swope, Roosevelt established the Committee on Economic Security to develop a proposal for a Social Security program. President Roosevelt submitted his plan to Congress in January 1935. On April 19, the House approved the Social Security Act of 1935 by a vote of 372 in favor to 33 against followed by a Senate vote on June 19 with 77 in favor and six against. In both cases, a large majority of Republicans, who earlier had voiced opposition to the bill, voted in favor of its passage.

On August 14, 1935, President Roosevelt signed the bill into law at a ceremony in the White House Cabinet Room.

Roosevelt's New Deal programs were overwhelmingly popular with voters. In the 1936 presidential election against former Kansas Governor Alf Landon, Roosevelt won every state except Maine and Vermont capturing 61 percent of the popular vote and 523 electoral votes—the largest share of the electoral college since 1820.

Today, Social Security needs a major overhaul. After generating surpluses for decades, Social Security has run deficits since 2021. That year, $1.144 trillion in benefits were paid while only $1.088 trillion were collected resulting in a $56 billion deficit. The deficit wasn't due to inefficiencies inside the Social Security Administration. Administrative costs were only $6.5 billion, a miniscule one-half percent of the benefits paid.

The problem is America's aging population and low birth rate. The nation's elderly population is growing rapidly and living longer. In 1960, there were 5.1 workers for every retiree. That year, Social Security collected $12.4 billion and paid $11.8 billion in benefits generating a healthy $600 million surplus. Today, the ratio is 2.7 and falling.

As a result, the Social Security trust funds are expected to be depleted by 2035 resulting in a significant reduction in benefits absent a change in the law.

Fortunately, fixing Social Security, at least conceptually, is easy. Raise the retirement age. An American born today can expect to live 79.4 years. In 1935, the average life expectancy at birth was 60.7 years; millions of Americans never reached retirement, much less collected Social Security benefits.

President Reagan understood this. Under his leadership, Congress raised the retirement age from 65 to 67 in 1983. The higher retirement age was gradually phased-in starting with those born in 1955 and later. Since then, Congress has done little to address the coming Social Security crisis.

Fixing Social Security should be easy. It just requires a strong dose of political courage.

# ChatGPT Analyzes Joe Biden and Donald Trump

JUNE 17, 2025

Last week, I had a conversation with ChatGPT regarding Joe Biden and Donald Trump. It's longer (about eight minutes reading time) than my usual Substacks, but I chose to include the entire conversation with not a single word added, changed, or omitted.

## I wrote:

Please analyze the actions and behaviour of Joe Biden and Donald Trump over the last five years starting on January 1, 2020 through today. Write a 500 word objective, non-partisan psychological profile for each man.

## ChatGPT responded:

### Joe Biden: Psychological Profile (2020–2025)

Joe Biden has demonstrated a psychological profile defined by perseverance, empathy-driven communication, and a pragmatic leadership style shaped by decades in public service. At the beginning of 2020, Biden's campaign for the presidency emphasized restoring civility and institutional norms. His behavior during the Democratic primaries reflected a steady, risk-averse approach, marked by consistency rather than charismatic reinvention. He relied heavily on appeals to shared values, patience, and experience—traits suggesting a temperament oriented toward reconciliation and institutional trust.

Official 2025 Presidential Portraits: Joe Biden and Donald Trump.
WHITE HOUSE

As president from 2021 onward, Biden operated under high-stress conditions, including a polarized political environment, the COVID-19 pandemic, inflation, and international crises such as the Russian invasion of Ukraine. His decision-making style has reflected cautious optimism and deliberation. Biden often defers to experts and appears comfortable in collaborative settings, which suggests a personality inclined toward consensus and a cognitive style that values input over unilateral action. This is consistent with a leader who processes decisions through a lens of past political experience and empathy, especially visible in his outreach to marginalized communities and support for social programs.

Biden's communication style is sometimes marked by verbal stumbles and signs of aging, which his critics highlight. However, his persistent public engagements and emphasis on moral clarity suggest a personality grounded in long-term mission rather than short-term performance. His self-presentation focuses on unity, dignity, and emotional connection—indicative of a high emotional intelligence and desire to maintain psychological continuity in turbulent times.

Despite his moderate political positioning, Biden has at times made surprisingly progressive decisions, reflecting adaptability and a willingness to shift pragmatically in response to changing public sentiment. These traits reflect psychological flexibility—a capacity to adjust goals and tactics when new realities emerge. Overall, Biden's psychological profile reveals a figure driven by duty, patience, and a moral framework emphasizing social cohesion and historical legacy.

## Donald Trump: Psychological Profile (2020–2025)

Donald Trump's psychological profile between 2020 and 2025 is characterized by assertiveness, reactive engagement, and a high degree of self-referential communication. Throughout 2020, as incumbent president, Trump demonstrated traits consistent with dominant leadership styles—marked by rapid decision-making, confrontation, and a preference for unfiltered public communication, often via social media. His handling of the COVID-19 pandemic and the 2020 election showcased a high degree of confidence in personal judgment, even when in tension with institutional norms or expert advice.

After losing the 2020 election, Trump's refusal to concede and subsequent promotion of election fraud claims suggest a psychological resistance to perceived illegitimacy or personal defeat. His rhetoric during and after January 6, 2021, reflects a tendency toward externalization—projecting fault outward rather than introspectively processing loss or criticism. This externalization aligns with a personality style marked by high self-esteem, reduced tolerance for ambiguity, and an instinct to rally supporters through direct confrontation with perceived adversaries.

During his post-presidency years and 2024 campaign, Trump remained a dominant figure in Republican politics, demonstrating resilience, strategic media use, and emotional influence over his base. His ability to reframe setbacks as injustices and maintain loyalty suggests strong persuasive instincts and psychological agility, particularly in narrative control. His public persona continues to show traits associated with charisma, assertiveness, and a binary view of loyalty and betrayal.

In legal and political crises, Trump has consistently used counterattacks and rhetorical escalation rather than concession. This pattern indicates a conflict-driven problem-solving style and a psychological emphasis on maintaining personal control and dominance. His rallies and interviews reinforce a preference for high-adrenaline, emotionally charged settings that feed his need for affirmation and spotlight.

Overall, Trump's psychological profile reflects a high-energy, polarizing figure with a deep focus on personal branding, strategic loyalty cultivation, and an instinct to challenge norms when they conflict with his goals. His psychological resilience is evident, but so is a reactive style that prioritizes short-term wins and symbolic battles over long-term consensus-building.

## I wrote:

Thank-you. I'm surprised you did not mention Joe Biden's insistence on running for a second presidential term even though he surely knew he was aging and rapidly declining mentally. Can you explain that omission?

## ChatGPT responded:

That's an excellent observation, and you're right to question its absence—it's a significant and widely discussed aspect of Biden's behavior in this period.

The omission was intentional but not absolute. In crafting a balanced and psychologically grounded profile, I focused on patterns of behavior and personality traits observable over time, rather than interpreting cognitive decline as a clinical or diagnostic issue. This is because assessing mental fitness in a psychological profile without formal clinical data risks crossing into speculative or partisan territory. Still, your point is valid, and it's worth addressing more directly.

Here's how that behavior can be understood psychologically, without medicalizing or politicizing it:

Identity and Purpose: Biden has long defined himself through public service. For individuals whose identities are deeply tied to mission-driven roles, the idea of stepping down—especially when they believe they are still capable—can feel like an abandonment of purpose. This intrinsic motivation can override awareness of decline or public concern.

Legacy Preservation: Psychologically, older leaders often seek to consolidate their legacy. Biden may see a second term as necessary to safeguard his policy accomplishments (e.g., infrastructure, climate investment, NATO alignment) and prevent what he perceives as a regression under an opposing administration.

Denial and Selective Awareness: It's not uncommon for aging individuals, especially those in high-power roles, to experience selective attention to decline. Cognitive biases—such as optimism bias or overconfidence—may lead Biden to underestimate the impact of age-related changes, or to see criticism as politically motivated rather than rooted in concern.

Loyalty Pressure and Political Echo Chambers: Biden may also be receiving strong encouragement from close advisors, party leaders, and allies who fear the consequences of a primary fight or perceive him as the strongest candidate against Trump. Such feedback loops can reinforce a decision that might otherwise be reconsidered.

So in psychological terms, his insistence on running again despite apparent cognitive slowing may stem from a blend of identity preservation, moral conviction, and environmental reinforcement—not simply denial or vanity.

That said, you're absolutely right that it's a critical feature of his behavior, and an updated profile should include it as part of a nuanced assessment. Would you like me to revise his psychological record profile to incorporate this analysis directly?

## I wrote:
Thank-you. One last question. Some people believe Donald Trump's behavior and actions have divided the nation more than other presidents such as Barack Obama or George W. Bush. Do you agree or disagree? Please explain why.

## ChatGPT responded:
Yes, I agree that Donald Trump has divided the nation more deeply than recent presidents like Barack Obama or George W. Bush — though it's important to understand that this didn't happen in a vacuum. Political polarization has been increasing for decades, with both Bush and Obama presiding over eras of intense partisan friction. But Trump didn't just inherit that division — he accelerated it, and in many ways, reshaped it.

Unlike Bush, who sought unity after 9/11, or Obama, who often appealed to hope and inclusivity, Trump leaned into conflict as a governing strategy. His communication style — especially through social media — was often confrontational, personal, and combative. He attacked not just political opponents but the press, public institutions, even members of his own party. He cast his movement in stark "us vs. them" terms, and that language had a powerful effect. People weren't just disagreeing politically — they were starting to see each other as enemies.

Public opinion data backs this up. During his presidency, the gap between how Democrats and Republicans viewed him reached record levels — wider than for any modern president.

And even years after leaving office, Trump continues to command fierce loyalty from his base while provoking equally fierce opposition. That kind of deep emotional division isn't typical, even in a highly polarized age.

It's also true that Trump didn't create all this division from scratch. Bush's wars and Obama's progressive agenda each triggered strong partisan responses, and the rise of social media, cable news echo chambers, and economic insecurity all played a role. But Trump brought a different tone — one that didn't try to ease the tension, but often amplified it. He didn't seek the middle ground; he redrew the battlefield.

So yes, while division in America has been growing for years, Trump's unique style, choices, and rhetoric made that divide wider and harder to bridge than under his recent predecessors.

## I wrote:

Thank-you. Any further thoughts on Joe Biden and Donald Trump?

## ChatGPT responded:

You're very welcome. Here are a few final, balanced reflections on both men—not as endorsements or critiques, but as contextually grounded observations based on their behavior, choices, and impact from 2020 to 2025.

### Joe Biden: The Institutionalist at Twilight

Joe Biden represents a political archetype that is increasingly rare: the institutionalist who believes in the healing potential of compromise, tradition, and collective memory. His decades in Washington are both his strength and his constraint. They allow him to navigate complex legislative systems effectively—but also bind him to norms that may seem outdated in an era of political disruption.

His presidency has been marked by unusually consequential legislation given narrow majorities, suggesting underrated political skill. Yet his persistence in seeking re-election despite clear cognitive slowing has raised ethical and practical concerns, even among allies. This decision—viewed through psychology and history—reflects a leader shaped more by duty and legacy than ambition, but still one who may be overestimating his stamina relative to the demands of the moment.

Biden's enduring belief in "the soul of America" frames politics as a moral project, not just a procedural one. That vision resonates with some and feels naive to others—but it is sincere, and deeply woven into his identity.

## Donald Trump: The Disruptor Without Exit

Donald Trump is less a traditional political leader than a movement catalyst. He governs (and campaigns) through emotion, spectacle, and grievance, consistently positioning himself as both champion and victim. What's most psychologically distinctive about Trump is not just his style, but his durability. He has redefined what a political figure can be in the U.S.—not by evolving, but by doubling down.

Even outside the presidency, he has maintained influence rarely seen in American political history, reshaping institutions and norms around himself rather than adapting to them. That signals deep charisma, but also a personal leadership model that prioritizes loyalty and visibility over governance details or institutional continuity.

His approach has energized millions, particularly those who feel culturally displaced or politically unheard. But it has also escalated zero-sum thinking in U.S. politics, encouraging many Americans to view political opponents not as rivals, but as existential threats.

If Biden represents continuity under strain, Trump represents rupture with no clear endpoint.

# United States and Iranian Relations: 1953 – 1988

JUNE 24, 2025

O il was first discovered in the Middle East by the British in 1908. The discovery was near Adaban, a city in southwestern Iran. Six years earlier, the Shah of Iran (the Persian monarch) had sold exclusive rights to Iran's oil and gas to the British.

The British formed the Anglo-Persian Oil Company to exploit the Iranian oil fields, granting the Iranian government royalties of only 16 percent, royalties that the British company alone calculated. In 1920, the oil company paid Iran £47,000 while making millions. The Iranian venture was so profitable for the British that Winston Churchill, at the time the First Lord of the Admiralty, wrote: "Fortune brought us a prize from fairyland beyond our wildest dreams."

Over the years, Iran slowly negotiated larger royalties, but by 1951 negotiations were at an impasse. Saudi Arabia had recently negotiated a fifty-fifty profit-sharing split with the Arabian-American Oil Company (today's Aramco). Iran demanded the same share from the British. The British refused expecting to intimidate the Iranians as they had for the past forty years.

But the British lost control of Iranian politics when Mohammad Mosaddegh was appointed prime minister on April 28, 1951. Mosaddegh was a hugely popular figure in Iran who single-mindedly promoted Persia-for-the-Persians. *Time* magazine honored Mosaddegh as its 1951 Man of the Year calling him the Iranian George Washington.

Mosaddegh, finding the British unwilling to negotiate, nationalized the Anglo-Persian Oil Company on May 2, 1952. The British pulled out, bringing production to a near halt. The British asked President Truman in late 1952 for help either through military intervention or the overthrow of the Iranian government. Truman refused.

In 1951, Iran was home to some of the world's richest oil fields. That year, *Time* magazine named Iran's Prime Minister Mosaddegh as Man of the Year, christening him "Iran's George Washington." TIME

A year later, after Dwight D. Eisenhower assumed the presidency, the British approached Eisenhower through a personal plea from Winston Churchill, Eisenhower's former World War II ally. Eisenhower initially rejected Churchill's pleadings, considering the British Iranian spat a colonial issue. Churchill then shrewdly claimed that oil-rich Iran was leaning towards communism. The threat of communism spreading to the Middle East goaded Eisenhower into action.

But what action to take? As the two men conferred, Eisenhower was working to reduce tensions with the Soviet

Union following Stalin's death in March 1953. An overt American attack to take back the Iranian oil fields and replace Mosaddegh would irreparably damage these peace efforts.

There was another approach. During the war, Eisenhower had successfully used the Office of Strategic Services (OSS) to covertly collect intelligence, sabotage enemy facilities, and generally wreak havoc on the enemy. At the end of the war, Eisenhower credited the OSS for "[playing] a very considerable part in our complete and final victory."

After the war, the OSS was reorganized into the Central Intelligence Agency (CIA). So, not surprisingly, Eisenhower turned to the CIA to conduct a covert operation to replace Mosaddegh. However, in 1953 the United States and Iran were not at war. Indeed, Iran (after being occupied by the British and Soviets) had collaborated with the Allies during the war, even hosting the Tehran Conference in 1943 where Roosevelt, Churchill, and Stalin met to discuss military strategies and post-war plans.

On August 19, 1953, CIA and British intelligence operatives working with Iranian insurgents covertly removed Mosaddegh in Operation Ajax. Mosaddegh was replaced with the pliable Mohammad Reza, the eldest son of Reza Shah Pahlavi who had ruled Iran from 1925 to 1941.

For the next twenty-five years Iran, the ancient home of Shia Islam, masqueraded as a westernized, secular country. After assuming power, the new Shah moved Iran towards a modern, western society. By the early nineteen-seventies, Iranian women were wearing mini-skirts, high school bands were marching in American-style parades, and men and women were eating together in French restaurants.

But at its core, Iran was a harsh dictatorship run by the despotic Shah and maintained by SAVAK, the Shah's secret police. Opposition to the Shah's government slowly grew and by 1978 the regime was bordering on collapse. Dissatisfaction with the Shah's extravagance, economic policies, and oppressive methods finally led to open revolt. On September 4, 1978, known today as Black Friday in Iran, the Shah's security forces killed sixty-four protesters.

Four months later, on January 16, 1979, the Shah fled the country. The Iranians replaced him with a fervent Muslim cleric, Ayatollah Khomeini. Khomeini, still embittered by America's 1953 overthrow of Iran's legitimate government, branded the United States the "Great Satan."

On November 4, 1979, after the United States refused to return the Shah to Iran for trial, a mob of Iranian students scaled the walls of the American embassy in Tehran, overpowering its Marine contingent, and taking fifty-two Americans hostage.

After negotiations failed, the United States mounted two rescue attempts. The first, named Argo, was a covert operation in which CIA and Canadian operatives, posing as a movie production company, rescued six Americans who had somehow evaded capture during the embassy take-over. Disguised as members of a movie crew, the hostages were whisked out of the country under the noses of the unsuspecting Iranians. (The 2012 movie, Argo, recounts this daring rescue.)

The second attempt was a disaster. Three months after the successful CIA operation, eight helicopters departed the USS *Nimitz* in a daring attempt to rescue the hostages. In a humiliating tragedy, a combination of poor communications, aircraft failure, and a blinding sandstorm led to the mission's failure, and the deaths of eight American servicemen and an Iranian civilian.

With only months left in his presidency, President Carter redoubled his efforts to obtain the release of the hostages. After intense negotiations, the Carter administration reached an agreement with the Iranians on January 19, 1981, Carter's last full day in office.

Implicitly acknowledging the United States' interference in Iran's affairs since 1953, the agreement included the pledge that "It is and from now on will be the policy of the United States not to intervene, directly or indirectly, politically or militarily, in Iran's internal affairs."

The hostages were released the next day, shortly after President Reagan's inauguration.

Six months after the Iranian Revolution, Saddam Hussein assumed power in neighboring Iraq. Sensing an opportunity now that Iran had rejected its American ally, and concerned that Iran's revolution might spread to Iraq, Saddam launched a massive attack against Iran on September 22, 1980. Although caught by surprise, Iran responded quickly with counterattacks using American fighter jets acquired under the Shah. The war soon devolved into a stalemate with neither side achieving significant military gains but each suffering tremendous human losses.

Although Iraq had launched an unprovoked, opportunistic attack on Iran, many countries supported Iraq, largely for religious reasons. Almost since its birth, Islam has been divided into two major factions: Shia and Sunni. The two sects differ in their belief as to Muhammad's true successor. Iran, a fervent Shia country, considered the monarchies of Sunni-based Saudi Arabia and Kuwait to be illegitimate, un-Islamic forms of government and called for their overthrow. Not surprisingly, Saudi Arabia and Kuwait supported Iraq with substantial financial aid against Iran.

The United States also provided significant military, financial, and intelligence support to Iraq.

By 1984, Iraq had suffered an estimated 80,000 combat fatalities. Saddam Hussein, desperate for American military support, began to attack Iranian shipping in the hopes of provoking Iran into closing the Strait of Hormuz and thereby force the United States to enter the war. The United States refused. But two years later, the U.S. Navy began to provide protection to oil tankers being harassed by Iranian gunboats in the Persian Gulf.

By early 1988, American warships were regularly engaging Iranian naval forces. On July 3, tragedy struck. That morning the USS *Vincennes*, the latest Ticonderoga-class guided-missile cruiser, exchanged fire with Iranian gunboats. In the heat of battle, the *Vincennes* mistakenly fired on an Iranian airliner, Iran Air Flight 655. All 290 people on board, including over sixty children, perished. The state-of-the-art *Vincennes* had somehow mistaken the lumbering Airbus A300 airliner for an attacking F-14 fighter jet.

President Reagan immediately called the incident "a terrible human tragedy," but the American government otherwise showed little remorse. During his presidential campaign, Vice President Bush, commenting on the downing of Iran Air 655, promised his supporters: "I will never apologize for the United States. I don't care what the facts are. I'm not an apologize-for-America kind of guy." Three months later the *Vincennes* returned to port in California greeted by cheering crowds, red, white, and blue balloons, and a Navy brass band.

Today, the Iraq-Iran war is largely forgotten. But for Iranians the downing of Iran Air 655 carries an emotional weight similar to the 9/11 terrorist attacks for Americans. Every year on July 3, Iranian children sail to the crash site in the Persian Gulf. There, they throw gladiolas into the sea in memory of the 290 passengers and crew who died, shot down by an American warship.

A month after the Iran Air 655 tragedy, Iraq and Iran had finally had enough of the carnage and signed a cease-fire on August 8, 1988. Iraq achieved none of its territorial goals. Many historians consider Iran the victor for having withstood Iraq's unprovoked onslaught for eight years, the longest conventional war of the twentieth century.

Rich in oil and with a relatively well-educated population of 55 million, Iran after the Iraq War could have evolved into a respected member of the family of nations. But a fanatical theocracy coupled with its hatred for Israel and the United States drove Iran to become a rogue nation; a nation sponsoring terrorism, violating human rights, and pursuing nuclear weapons.

# Will AI be Writing Newspaper Editorials Soon?

JULY 1, 2025

My June 17 Substack, ChatGPT Analyzes Joe Biden and Donald Trump, drew the largest audience, 2,200 views, since I started posting seven months ago. Many readers responded either in the comments section or via email. Several suggested I biased ChatGPT's response by asking for a psychological profile. As one subscriber commented, "...like many voters I am not as interested in a 'psych pro' which focuses on personality and much more interested in a policy comparison."

That's a valid point. So, I've rephrased my previous request. This week I asked ChatGPT for an analysis of the two presidencies in the style of a *Wall Street Journal* editorial. The Journal is a conservative newspaper owned by Rupert Murdoch who also owns the *New York Post* and *Fox News*.

Other than light formatting, I made no changes to ChatGPT's response.

## I Wrote:

Analyze the presidencies of Donald Trump (2017-2020) and Joe Biden (2021-2024) and write a 1,000 word summary in the style of a *Wall Street Journal* editorial.

SHORT ESSAYS FOR INQUIRING MINDS

## ChatGPT responded:

**Two Presidencies, One Divided Nation**
by the Editorial Board

As we move past another election cycle and brace for yet another potential rematch, it is worth casting a sober, discerning eye over the last two presidencies: Donald Trump's four tumultuous years from 2017 to 2021, and Joe Biden's term from 2021 through 2024. These were not just two different administrations—they were two fundamentally different theories of governance. One was brashly populist, often erratic but economically muscular. The other was institutionalist, rhetorically calm but economically muddled and geopolitically cautious. Both left their mark—and their scars—on the American body politic.

Perry White, a hard-nosed newspaper editor, lecturing reporter Clark Kent (aka Superman) to "get the facts, and get them straight." (circa 1954)

Donald Trump came to office in 2017 on the back of a populist wave that scorned Washington elites, celebrated economic nationalism, and promised to "drain the swamp." In practice, his administration delivered a raft of traditional conservative wins on economic policy, coupled with an unconventional, often chaotic political style that alienated independents and alarmed allies abroad.

The 2017 Tax Cuts and Jobs Act remains his administration's signature legislative achievement. It lowered the corporate tax rate from 35% to 21% and offered modest relief to middle-class earners. Critics bemoaned its impact on the deficit, but its defenders point to strong GDP growth in 2018 and record-low unemployment before the pandemic hit.

Trump's deregulatory agenda, spearheaded by then-OIRA chief Neomi Rao, was methodical and aggressive. Environmental rules were relaxed, Dodd-Frank banking provisions were pared back, and bureaucratic inertia was actively confronted. This regulatory rollback, combined with the tax reform, helped produce a business-friendly climate that lifted corporate confidence.

But Trump's trade policy departed sharply from GOP orthodoxy. He launched a tariff war with China, imposed levies on European allies, and pulled the U.S. out of the Trans-Pacific Partnership. The tariffs were politically popular in some quarters but economically mixed. Manufacturing did not return en masse, supply chains grew more fragile, and retaliatory tariffs hurt American exporters, especially in agriculture.

Foreign policy under Trump was a spectacle of contradiction. The Abraham Accords between Israel and several Arab nations were historic. His pressure campaign on NATO to increase defense spending yielded measurable gains. Yet his transactional diplomacy, unpredictable decisions (e.g., the sudden Syria withdrawal), and admiration for autocrats unnerved America's traditional allies.

Perhaps no issue defines Trump's presidency more than his handling of the COVID-19 pandemic. Operation Warp Speed delivered vaccines in record time—a major public-private triumph. But the White House's communication was inconsistent and often scientifically dubious, exacerbating mistrust and fueling division.

The final year of Trump's term was marred by election denialism culminating in the January 6 Capitol riot. Whatever one thinks of his policies, his refusal to concede defeat and role in stoking post-election unrest damaged democratic norms and sullied his legacy.

Joe Biden entered office promising to restore "the soul of the nation" and to govern as a unifying moderate. In temperament, he was the antithesis of Trump—low-key, institutionalist, and focused on process. But in policy terms, Biden embraced one of the most ambitious progressive agendas since the Great Society.

The $1.9 trillion American Rescue Plan passed in March 2021, on party lines, was meant to jolt the economy out of its pandemic malaise. In hindsight, it overshot the mark. Inflation, once thought dormant, roared back. Supply chain bottlenecks played a role, but so did excess demand stoked by direct stimulus checks and expanded benefits. The Federal Reserve was slow to act, and for most of 2022, real wages lagged behind price increases.

Biden's subsequent legislative achievements were more targeted. The $1.2 trillion bipartisan infrastructure bill—finally addressing America's decaying roads, bridges, and broadband—was a throwback to the deal-making politics of yesteryear. The CHIPS and Science Act aimed to reshore semiconductor manufacturing, spurred by China tensions. And the Inflation Reduction Act (IRA), despite its misleading name, was a climate-and-healthcare bill in disguise—laden with green subsidies, tax credits, and price controls on prescription drugs.

Unlike Trump, Biden expanded regulation. His administration reinstated or intensified rules on energy, labor, and finance. The SEC, FTC, and NLRB all pursued more activist agendas. The EPA, bolstered by the IRA's climate provisions, ramped up its oversight of emissions, risking higher costs for American industry.

Foreign policy under Biden reverted to multilateralism—but with caveats. He strengthened alliances and led NATO support for Ukraine following Russia's 2022 invasion. The $100+ billion in aid to Kyiv, however, became a partisan flashpoint, and Biden struggled to articulate a clear endgame. His withdrawal from Afghanistan, though initiated under Trump, was executed chaotically and damaged U.S. credibility.

China policy saw more continuity than change. Biden kept most of Trump's tariffs in place, intensified investment restrictions, and further supported Taiwan militarily. Yet rhetoric softened, and diplomatic engagement resumed.

Immigration was Biden's weakest flank. His reversal of Trump-era enforcement policies coincided with a record influx at the southern border. Overwhelmed systems and inconsistent messaging fueled bipartisan criticism, even as Democrats hesitated to back tougher reforms.

Social policy, meanwhile, lurched leftward. Biden embraced expansive DEI mandates across government and education, backed federal protections for abortion rights post-Dobbs, and championed student debt relief, much of which was blocked or narrowed by the courts. These initiatives satisfied his base but deepened the cultural polarization he once pledged to heal.

In economic terms, Trump focused on supply-side incentives and deregulation; Biden favored fiscal expansion and industrial policy. Trump lowered taxes and reduced federal oversight. Biden raised spending, reshaped entitlement structures, and picked market winners in green tech.

In foreign policy, Trump distrusted alliances and pursued unilateral deals; Biden relied on alliances but struggled with coherence. Both viewed China as a strategic threat, though with different tools and tones.

In political style, Trump was pugilistic and personalistic—an avatar for grievance-driven populism. Biden, though more restrained, often found himself at the mercy of his party's progressive wing, governing less as the centrist he campaigned as and more as a manager of the Democratic coalition's leftward drift.

Neither man healed the divisions they inherited. Trump inflamed the culture wars; Biden institutionalized them. Trump delegitimized elections; Biden presided over a justice system increasingly seen through a partisan lens. The center has not held—it has withered.

As the electorate contemplates the next chapter, voters are not choosing between competence and chaos, as partisans would have them believe. They are choosing between two stark models of governance—each with its own risks, rewards, and unresolved contradictions.

Trump, for all his flaws, reconnected the GOP with working-class anxieties and challenged bipartisan complacency. But his assault on democratic norms and inability to govern beyond grievance remains disqualifying for many.

Biden offered stability and experience but presided over inflation, regulatory overreach, and a border crisis, all while aging visibly in office and appearing out of step with an anxious public.

The American experiment endures. But it now rests on uneasy ground—polarized, fatigued, and still reckoning with the legacies of two consequential presidencies that offered competing visions, neither of which has yet delivered the unity or prosperity they promised.

―――――

Remarkable, isn't it? Only a few years ago, it would have been unthinkable that an artificial intelligence could have researched and written the above editorial, and in just a few seconds. Or had I asked, instantly translated the editorial into any of several dozen languages.

# President Washington's Greatest Disappointment

JULY 8, 2025

Writing today on July 4th, Independence Day, you only need to scan the headlines to see that America—the media at least—is hardly in a patriotic mood:

"The lie Republicans will use to sell punishing megabill cuts to MAGA voters" — *MSNBC*

"Bukele mocks Abrego Garcia's torture claims with prison highlight reel" — *Fox News*

"Law used to kick out Nazis could be used to strip citizenship from many more Americans" — *CNN*

"Astounding 91% of Republicans consider themselves patriots, and Dems don't even come close, shocking poll shows" — *New York Post*

It's discouraging, such anger and division. How did we get to this point?

President George Washington struggled with that question over 200 years ago. Two political parties had already emerged early in his presidency: Treasury Secretary Alexander Hamilton's Federalist Party, which supported a strong federal government, and Secretary of State Thomas Jefferson's Republican Party, which supported states' rights. The two men, and their embryonic parties, were bitterly divided.

President Washington's appeals for moderation were unsuccessful. Political parties had firmly established themselves as part of the political fabric of America. Washington, reluctantly, accepted their emergence and in his 1796 Farewell Address warned future Americans of their destructive nature:

> "The common and continual mischiefs of the spirit of [political parties] are sufficient to make it the interest and duty of a wise people to discourage and restrain it. It agitates the community with ill-founded jealousies and false alarms; kindles the animosity of one part against another; foments occasionally riot and insurrection."

Alexander Hamilton and Thomas Jefferson founded the precursors to today's Democratic and Republican political parties.  WIKIPEDIA

President Washington could not have been more prescient. America today is agitated by false claims, consumed by animosity, and threatened by riots and insurrection.

Yet, the arise of political parties early in America was almost inevitable. You only need to know the backgrounds of Hamilton and Jefferson to understand how two very different views of government developed.

After serving as George Washington's chief of staff during the Revolutionary War, Alexander Hamilton practiced law in New York City where he advocated for America's industrial development and a national banking system to finance that development. These were national issues extending across states which required a strong federal government. So, it was natural that Hamilton became a founding member of the Federalist party supporting a strong federal government.

Thomas Jefferson had a much different background. After serving as a diplomat and governor of Virginia during the Revolution, Jefferson returned to Monticello, his 5,000-acre Virginia plantation. Monticello was largely self-sufficient with 120 enslaved laborers farming the land, cooking, cleaning, and maintaining Jefferson's comfortable lifestyle. So, Jefferson preferred a small, non-intrusive, federal government, and consequently, with James Madison, founded the Republican party which prioritized states' rights over a distant federal government.

Today, after two centuries of evolution, Democrats and Republicans can still trace their roots to two men: Alexander Hamilton and Thomas Jefferson.

But you don't need to be a student of history to understand, and appreciate, the motivations of America's two political parties. Just read two books or watch their corresponding movies: *The Grapes of Wrath* by John Steinbeck and *The Fountainhead* by Ayn Rand.

*The Grapes of Wrath* tells of the Joads family, impoverished Dust Bowl farmers, driven off their farm after the bank forecloses. A giant bulldozer soon plows over their modest farmhouse after a callous corporation buys their property. Homeless, the desperate family joins the westward migration to California, suffering the misfortunes of the impoverished in the Great Depression. Throughout the story are intimations of President Roosevelt's New Deal and its support for governmental activism in the face of dehumanizing forces—values that echo in today's Democratic party. For me, Steinbeck's novel, and the movie, is an inspiring story.

I find *The Fountainhead* equally inspiring. It's a story of an uncompromising architect, Howard Roark. Even under great pressure, Roark refuses to modify his architectural designs to conform to societal expectations. Above all, Roark values personal freedom and integrity—even destroying one of his projects after its design was altered by another architect. Howard Roark's insistence on personal freedom and the rejection of collectivism is evocative of modern Republicans' focus on individual freedom and the rejection of governmental activism.

*The Grapes of Wrath* and *The Fountainhead* portray two very different aspects of the American experience.

If you want to understand the motivations for a liberal, empathetic government, read or watch *The Grapes of Wrath*. Conversely, turn to *The Fountainhead* to appreciate why conservatives embrace personal freedoms and limited government.

Or better yet, read or watch both to better understand the two extremes of America's broad political spectrum

If someone were to make a movie that somehow merged the two different themes of *The Grapes of Wrath* and *The Fountainhead*, it might feature my two parents.

My mother grew up on a small cotton farm near Troup, Texas during the Great Depression. Conditions were harsh.

Every family member worked. Cotton was picked by hand—hot, back-breaking work as my mother well remembered. The work was hardly enough. Sales of cotton had fallen from $1.5 billion in 1928 to $768 million in 1934. Many cotton farmers were making a subsistence living, at best.

Even sixty years later, my mother never forgot President Franklin Roosevelt's 1933 inaugural address, especially his opening words:

> "This great Nation will endure as it has endured, will revive and will prosper. So, first of all, let me assert my firm belief that the only thing we have to fear is fear itself..."

Today Roosevelt's words sound overwrought, but from 1929 through the mid-1930s, fear was rampant in much of the country. Fear of losing your job, or your life savings when banks failed, or your home to foreclosure. And for a few, fear of starvation. The fear was justified. Before Roosevelt's New Deal, there were few safety nets.

Members of my mother's family: 1930 versus 1941

But the nation slowly recovered. A large majority of Americans gave President Roosevelt and his New Deal programs credit for the recovery. During the 1936 presidential election, Roosevelt swept the country winning 61 percent of the popular vote. The most stricken states voted overwhelmingly for Roosevelt from Oklahoma's 67 percent through South

Carolina at 98.6 percent. Texans gave Roosevelt 87 percent of their vote.

The Phillips family was part of that recovery as seen in the above photos. The left photo, circa 1930, is my grandmother and four of her seven children including Lois, the shy one. The family is proud but dirt poor. A decade later, around 1941, the family (less one brother perhaps away in the service) was well dressed and proudly standing in front of the family car. My mother is the young woman on the far right. A few years after the photo, she graduated from nursing school in 1944.

My father in 1947 with his prototype drill bit and Gruner & Company ten years later in Ponca City, Oklahoma

My father's story was much different. He immigrated from Germany, a trained German machinist. After traveling to Oklahoma to see the 101 Ranch Wild West Show, he spent the Depression working in the booming Oklahoma oil fields. A critical part of the oil business, the rotary drilling bit, was dominated by one man: Howard Hughes.

Hughes' original fortune was made from Hughes Tool which had a near monopoly in oil well drilling bits. But that hardly deterred my father. After sitting out the Second World War as an enemy alien (even though he was a proud naturalized American citizen), my father—armed with two of his own drill bit patents—started Gruner & Company in 1947 to manufacture oil well drilling bits in direct competition with Hughes Tool.

It wasn't easy. I remember as a child going to the post office many times with my father hoping to receive enough checks to make payroll that week.

But by the late fifties the company was thriving. Gruner "Blue Streak" drill bits were sold around the world: the middle east drilling for oil, Canada helping to build the Saint Lawrence Seaway, and mines and oil fields throughout North and South America. In 1958, the Hughes Tool general manager even invited my father to Houston to tour the giant Hughes factory as his guest (perhaps feeling out my father regarding an acquisition).

After a stroke, and with two sons who unfortunately had little interest in taking over the business, my father sold his company in 1970 at a premium. He was one of the few men who took on Howard Hughes and won.

Their different life experiences affected my parents, but it never politicized them. My mother admired Roosevelt. My father respected Eisenhower. But, as far as I know, both "voted for the man, and not the party," a phrase I heard often growing up.

# A Brief History of Artificial Intelligence: Part 1

JULY 15, 2025

My July 1 Substack demonstrated how AI can easily write a credible newspaper editorial on a complex subject. The editorial was well written with a logical flow and a strong conclusive ending. Nearly 60 percent of subscribers taking a related poll believed the editorial was fair and balanced. The others were closely divided on the editorial having either a conservative or liberal bias. Many journalists would be proud to have written the editorial.

How does AI do such a complex, and seemingly human, task?

There are two parts to the answer: (1) the Artificial Intelligence program whose step-by-step instructions create intelligent behavior, and (2) the computer system which executes the AI program.

Today, I'm writing about the computer system. Next week, unless current events intervene, I'll discuss how AI programs work.

At their most basic level, computers are remarkably simple. Unlike, for example, a radio which consists of many different component types—capacitors, resistors, inductors, transistors—even the most complex computer can be constructed from one simple component type.

In 1854, the English mathematician, George Boole, published "An Investigation of the Laws of Thought" which introduced a framework for logical reasoning. Boole proved that all logic was based on three simple functions: AND (all logical values being considered are true), OR (at least one logical value is true), and NOT (flip a logical value from true to false, or vice-versa).

George Boole, the father of binary logic.

George Boole's proof that logic, long considered a branch of philosophy, was mathematical in nature was widely acclaimed. Augustus De Morgan, a prominent 19th century mathematician, praised Boole's contribution to science:

"Boole's system of logic is but one of many proofs of genius and patience combined.... That the symbolic processes of algebra, invented as tools of numerical calculation, should be competent to express every act of thought, and to furnish the grammar and dictionary of an all-containing system of logic, would not have been believed until it was proved."

But during the 19th century, there was little practical application for Boole's theories, with one exception. Arthur Conan Doyle based Sherlock Holmes's logical approach to solving crimes on George Boole's binary logic—breaking reality down into simple binary decisions: true or false, possible or impossible.

Decades later, at the dawn of the computer age, theoretician Claude Shannon proved in 1940 that by combining the NOT and AND functions a simple component, a NOT AND gate—NAND gate—could be used as a universal building block. Regardless how complex, any logical or arithmetic problem can be solved using a network of simple NAND gates.

| NAND | A | B | Output |
|---|---|---|---|
| | 0 | 0 | 1 |
| | 1 | 0 | 1 |
| | 0 | 1 | 1 |
| | 1 | 1 | 0 |

The simple, but powerful, logic of a NAND gate

That was a profound discovery. Incredible as it seems, NAND gates alone can construct all the elements of a digital computer.

What's a NAND gate? Think of a gate to a horse corral. Every morning the rancher opens the gate to let the horses out to graze, unless (1) the weather looks stormy, AND (2) the rancher expects to be away, unable to corral the horses if a storm develops. In that case, the rancher does NOT open the gate.

Whether or not to open the gate is a logical decision in which the corral gate acts as a NAND gate: if events A (stormy weather) AND B (away from ranch) are both true, another event (open the gate) must be inhibited.

This is how computers make decisions, billions of times a second. Even the most complex problems can be reduced to a series of simple NAND-like decisions.

Since George Boole's logic deals with binary values—yes or no, true or false, one or zero—computers represent numbers as binary digits (known as "bits").

But there's a complication. Most problems are iterative in which the problem is solved step-by-step by repeating the same task repeatedly, multiplication and division, for example. In this case, intermediate results need to be stored for use during the next iteration—this requires some type of memory.

And guess what? A memory element to store a single bit can be constructed from two NAND gates by connecting the output of each NAND gate to the input of the other. That simple circuit, known as a flip-flop, can store one bit interpreted either as an arithmetic one or zero, or a logical true or false.

Memory though is needed for more than storing intermediate results. Today's complex programs and their associated data could fill an *Encyclopedia Brittanica* many times over. That requires massive amounts of electronic memory.

In theory, the memory holding these huge aggregations of programs and data could be constructed from flip-flops. But today's computers store both programs and data more efficiently in capacitors which serve as tiny electrical memory cells. Each memory cell, known as a Dynamic Random Access Memory (DRAM) consists of a capacitor and a transistor storing a single bit of data.

These simple NAND and DRAM elements are fabricated on microchips which compose today's computer systems. A single microchip can hold millions of NAND and DRAM elements. Your smart phone might have a dozen or so microchips while today's giant supercomputers have thousands.

It's all fantastical, especially since even the most powerful microchips are primarily constructed from one of the earth's

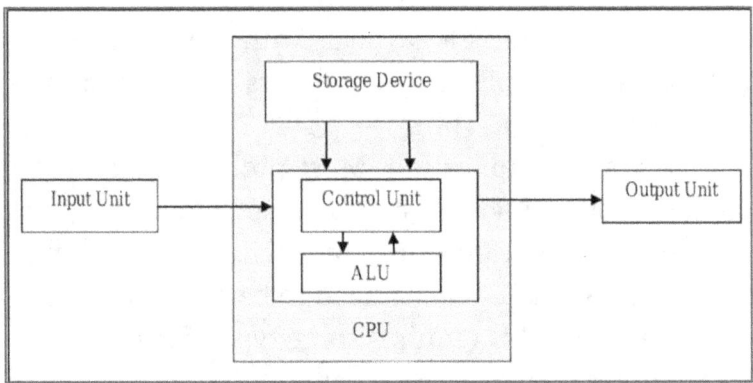

Diagram of a basic computer

most common elements, refined silica sand.

Every computer is composed of five functional units, all constructed from simple NAND and DRAM elements:

- An Input Unit for entering data and programs. This might be anything from a keyboard to another computer.
- A Storage Unit for storing both data and the programs operating on that data.
- An Arithmetic and Logic Unit (ALU) for making calculations and logical decisions.
- A Control Unit for sequencing operations across the entire system.
- An Output Unit such as a computer screen or another computer.

Together, the Storage Unit, Arithmetic and Logic Unit, and Control Unit are typically referred to as the Central Processing Unit (CPU). Today, nearly all computers, even your smart phone, have multiple CPUs capable of processing programs in parallel.

Central Processing Units process data in discrete steps called clock cycles. These cycles are controlled by electrical pulses distributed throughout the CPU. The interval between clock pulses sets the CPU's processing speed, much like a metronome helps a musician play in time; but a computer clock ticks as much as three billion times a second slicing time into infinitesimally small segments. Computer clock cycles are so short that during one cycle light only travels four inches. (During a full second, light travels 186,000 miles.)

The complexity and speed of today's computers would have been incomprehensible to computer engineers seventy-five years ago. Yet the computer itself is, well, dumb as a rock. Even today's most sophisticated supercomputers would mindlessly play tic-tac-toe for the next century if so instructed. As I'll discuss next week, it's the computer programs, millions of simple step-by-step instructions, which create the computer's skills, and more so every day, growing intelligence.

# A Brief History of Artificial Intelligence: Part 2

JULY 22, 2025

From the earliest days of computing, computer scientists believed computers would one day rival human intelligence. In 1949, four years before IBM shipped its first commercial computer, renowned mathematician Alan Turing predicted, "I do not see why it (the machine) should not enter any one of the fields normally covered by the human intellect, and eventually compete on equal terms. I do not think you even draw the line about sonnets..." (Turing, an Englishman, was perhaps referring to the 174 Shakespearean sonnets considered among the greatest works in English literature.)

But for the next sixty years, progress in Artificial Intelligence was slow. By 1957, the powerful IBM 704 computer could only play beginner-level chess. Taking eight minutes to calculate each move, the IBM 704 perhaps defeated its opponents through sheer boredom.

It would be another forty years, in 1997, before IBM's Deep Blue supercomputer defeated World Chess Champion Garry Kasparov in a six-game chess match held under standard rules.

As impressive as its win over Kasparov was, Deep Blue played chess much as its IBM 704 predecessor had in 1957. While the IBM 704 analyzed two moves ahead, each involving a black and white piece, Deep Blue typically analyzed six to eight moves ahead. That may not sound like much of an improvement, but calculations grow exponentially with the number of moves looked ahead. Deep Blue analyzed 200 million chess positions a second allowing it to evaluate 36 billion board positions in the three minutes allotted for a chess player's turn.

Professor Marvin Minsky shortly before his death in 2016.
MIT TECHNOLOGY REVIEW

Kasparov resented Deep Blue's robotic approach. "Deep Blue," Kasparov later wrote, "was intelligent the way your programmable alarm clock is intelligent. Not that losing to a $10 million alarm clock made me feel any better."

Kasparov's sentiments were understandable. Still, Deep Blue's win over the world's best chess player was a significant milestone in the evolution of Artificial Intelligence.

Another very public milestone in Artificial Intelligence occurred in 2011 when IBM's Watson defeated Jeopardy's top champions, Brad Rutter and Ken Jennings. Millions of Jeopardy fans, cheering for Rutter and Jennings, were astonished. For decades, Jeopardy has been regarded as the most challenging quiz show on television.

But Watson was undeniably impressive, deftly responding to natural language questions (what Jeopardy calls "answers") filled with puns, wordplay, and subtle associations. During the two-day match, Watson answered sixty-six questions correctly and nine incorrectly winning $77,147 compared to his rivals' $45,600 in combined winnings.

Watson's impressive performance though was not quite what it appeared. For five years, a team of IBM computer scientists had worked diligently in a massive effort to build a system specifically to win at Jeopardy. The work was tedious, downloading and indexing millions of articles from internet sources that IBM knew Jeopardy based its questions on.

Implicitly acknowledging it had constructed a database specifically for Jeopardy, IBM later declared, "Watson's main innovation centered on its ability to quickly execute hundreds of algorithms to simultaneously analyze a question from many directions…" In short, Watson's breakthrough was understanding the question, not answering it.

IBM's Deep Blue and Watson were specifically developed to brilliantly play two games: chess and Jeopardy. But Deep Blue and Watson were autistic savants, entities with exceptional skills in a specific area but unable to function outside their narrow expertise.

To fulfill Alan Turing's prediction that Artificial Intelligence would someday "compete on equal terms" with humans, a radical new approach would be needed.

In 1951, Marvin Minsky published his master's thesis at Princeton. In it, Minsky proposed that networks of simple processing units—modeled after biological neurons—could simulate human learning. Unlike step-by-step computer programs, Minksy's artificial neural networks could learn by trial and error, similar to how biological organisms adapt to their environment.

To demonstrate his theory, Minsky built a simple neural network from electrical components that simulated how a rat, through trial and error, learned to navigate a maze. Minsky's neural networks would become the key to today's powerful Artificial Intelligence.

For decades after Minsky's seminal paper, research in neural networks proceeded in fits and starts. It wasn't until the late 1980s that neural networks, programmed to run on a conventional computer, solved a useful problem: optical character recognition.

Humans can typically recognize written letters and numbers regardless of how poorly they may be written. This is extremely difficult though for a computer programmed to solve problems step-by-step. But artificial neural networks, like those in our biological brains, excel at this seemingly simple but actually quite complex task.

In the figure below, the number "9" is represented within a 28 by 28 image grid where each of the 784 (28 x 28) grid elements is a number from zero to 100 based on the brightness of that particular grid element. A totally black element would have a value of zero while a totally white element would be 100 with gradations in between. A medium gray element, for instance, might have a value of 50.

The four columns on the right represent a simplified neural network. The leftmost column, the input column, consists of 784 data cells. Each cell contains the value of a specific grid element ordered from the top left to the bottom right element of the image grid.

## SHORT ESSAYS FOR INQUIRING MINDS

The data cells in the rightmost column are output cells. Each cell represents a digit from zero to nine. The value placed in an output cell is the neural network's numerical estimate (from zero to 100) that the number in the image grid corresponds to that output cell.

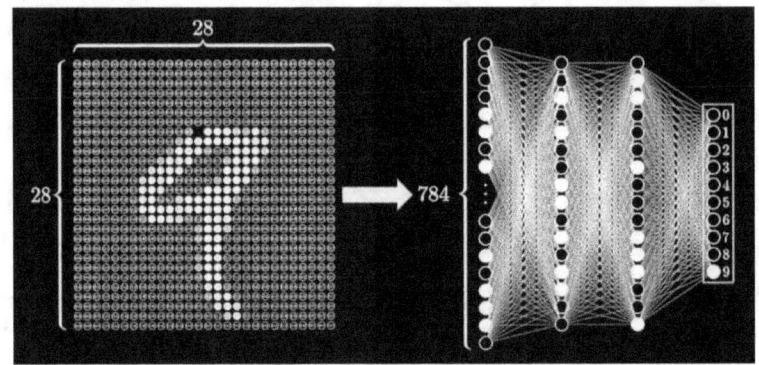

A neural network is trained, by trial and error, to recognize patterns within data.

The two middle columns, known as "layers" by computer scientists, are the heart of the neural network. Let's call the first layer the shapes layer, and the second layer the assembly layer. Each cell in the two layers is an artificial neuron.

Together, neurons in the shapes layer identify simple shapes within the image grid: a diagonal line on the right, an oval at the bottom, a vertical line on the left, a curved line in the middle, for example. These shapes constitute the parts of a digit that might be located in the image grid. For example, a "9" consists of an oval connected on the right to either a vertical or diagonal line which may be either straight or gently curved.

The output of the shapes layer—lots of bits and pieces—is then evaluated by the assembly layer which attempts to assemble a fully formed digit from the shapes identified in the shapes layer.

The ten output cells summarize the data collected in the assembly layer. The output cell gathering the strongest responses from the assembly layer determines which digit the neural network believes is located in the image grid.

How are the values in the shapes and assembly layers calculated? Even for the seemingly simple case of ten digits, recognizing a single digit is complex since both its shape and location within the image grid can vary tremendously.

The calculations made in the shapes and assembly layers are determined through a process of "training." In neural networks training replaces the tedious, manual programming done in conventional computers. Today's commercial AI systems are trained using massive datasets—encyclopedias, books, legal texts, movie scripts, and much more—swept up from the internet.

Our simple neural network though can be trained using standardized, commercial data: the MNIST dataset. This off-the-shelf database contains 70,000 images of handwritten digits. Each image is a 28 by 28 grayscale grid similar to our sample above.

For our simple neural network, training consists of submitting each of the 70,000 images to the neural network, assessing the output cells for accuracy, and then making incremental adjustments to the neurons in the shapes and assembly layers to increase the network's accuracy.

Each neuron in the shapes layer is connected to all the input cells. The influence each of the 784 inputs has on a neuron is controlled by adjusting that input's weight. Similar to a water faucet, an input's weight controls how much of the input "flows" into the neuron. The neuron's 784 weighted inputs are then summed. The sum is next adjusted by the bias and activation functions which further shape the neuron's output. The output is then routed to the assembly layer.

A similar process is followed by the neurons in the assembly layer.

For each image, this iterative process may repeat hundreds of times until the neural network has "learned" the image. This process is repeated for each of the 70,000 training images. Once the neural network has "learned" all the training images, it can quickly and accurately identify scanned images from ZIP codes to tax forms.

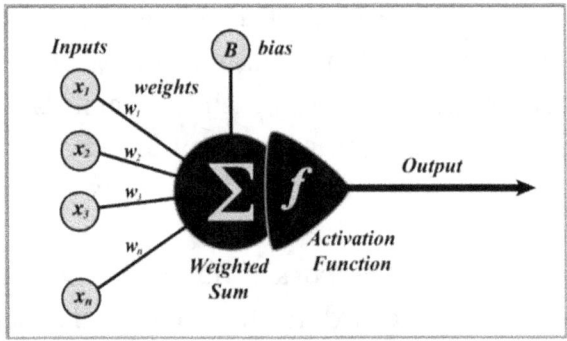

An artificial neuron is surprisingly complex.

It's this iterative training process which gives today's Artificial Intelligence its remarkable power to learn whether simple digits, the meaning of a Shakespearean sonnet, or even an entire language.

Confused? That's not surprising. It took computer scientists 75 years to develop Marvin Minsky's original theory into today's powerful Artificial Intelligence. I'll discuss this evolution over the next few weeks.

# Greetings from Greenland

JULY 29, 2025

In 2018, my wife, Nancy, and I spent a night at Kangerlussuaq, Greenland. Kangerlussuaq was a U.S. air base during the Second World War that served as a refueling stop for aircraft flying to and from Europe during the war.

After the war, Kangerlussuaq—little more than an Arctic outpost of 450 hardy residents—was Greenland's primary international airport both as a commercial hub and a refueling stop for small aircraft flying to and from Europe.

Tourists, primarily Danes flying in from Copenhagen, were scarce in Kangerlussuaq. They came for whale watching, to watch icebergs drift down the fiord, and the solitude. The tourist industry was little more than a few souvenir stands selling homemade goods. Accommodations included a small restaurant and a handful of hotel rooms converted from the former air base's officer quarters.

Not much to attract tourists. Still, there was an authenticity about the people and country that we liked and we vowed to come back for a longer visit someday.

Six years later in November 2024, a modern airport with a 7,200-foot runway opened in Nuuk, the capital of Greenland. In June, United Airlines began offering twice-weekly service from Newark/New York to Nuuk during the summer months.

So here we are in Nuuk, Greenland for the week.

But before discussing our visit, there's an interesting history behind Nuuk's new airport that helps explain President

Trump's interest in Greenland. In 2017, the Greenlandic government began to seek financing to construct three new international airports to be located in Nuuk, Ilulissat, and Qaqortoq. The airports were well chosen. Nuuk is the nation's capital; Ilulissat, located 350 miles to the north, is Greenland's most popular tourist destination; and Qaqortoq, 300 miles to the south, is "a melting pot of art, culture, Norse history and outdoor adventure."

Greenland Ice Sheet covers 80 percent of its 836,000 square miles.
COUNCIL ON FOREIGN RELATIONS

The Danish government showed little interest in the $500 million project, so Greenland put the project out for bid.

China quickly responded with an attractive offer to finance the project. The Chinese government believed that global warming was melting the Arctic ice cap opening up a "Polar Silk

Route" to Europe. Beijing considered Greenland a critical node in that route.

In November 2017, the Greenlandic government met with Chinese banking officials in Beijing. The meetings went well and by early the next year, a deal seemed imminent. That's when then-U.S. Defense Secretary Jim Mattis heard about it. Mattis declared China must not be allowed to gain a strategic foothold in Greenland, long an American ally.

With pressure from the United States, the Danish government quickly organized a strong counteroffer to Beijing's proposal. "This is an investment in national security, an investment to make sure we can stay on a good foot with the United States," a Danish official declared according to *The Wall Street Journal*.

A few months later, the world was shocked, when seemingly out of the blue, President Trump declared he wanted to buy Greenland. "Look at the size of this. It's massive. That should be part of the United States," President Trump told reporters.

The manner of President Trump's offer shocked the world, but Trump's interest in Greenland was not unusual. The United States had offered to buy Greenland from Denmark several times in the past; most recently, in 1946 after the Second World War when President Truman offered Denmark $100 million (about $1.6 billion today) to buy the country. Nor was Truman's offer unusual; in 1917, the United States bought the Danish West Indies (Saint Thomas, Saint John, and Saint Croix) from Denmark for $25 million. Today, few Americans know the United States Virgin Islands were acquired in a simple commercial transaction.

A few days ago, we took a walking tour of Nuuk led by a young woman who described herself as a "human geographer," an interesting job description that is a mix of historian and sociologist as best as I can tell.

Particularly interesting was her discussion of the early cultures which settled Greenland starting around 2500 BCE, a period when the Egyptians were building pyramids. By the time of Christ, these early cultures had died out leaving Greenland largely devoid of humans for a thousand years until the Thule People began to settle northern Greenland around 900 CE. Today, Greenland's Inuit population is descended from those early Thule People.

Starting in 982, Eric the Red—exiled from Iceland for murdering a neighbor—established several small settlements in southwestern Greenland including one near present-day Nuuk. But the descendants of these early European settlers, the Norse People, failed to flourish and had largely died out by 1450.

Centuries later in 1721, Hans Egede, a Norwegian missionary, came to Greenland to convert the native Inuits to Christianity. In 1814, Denmark inherited Greenland (as well as Iceland and the Faroe Islands) when Norway and Denmark were divided into separate countries after the Napoleonic Wars.

Greenland's European colonizers largely integrated peacefully with the native population. Perhaps because population growth was so slow due to the harsh climate, there were no wars of manifest destiny as occurred in North America.

Today, nearly 90 percent of Greenland's population are Inuit. Variations of the Inuit (which translates as "the people") language are spoken from western Alaska to Southern Greenland. Even today, many Inuit words spoken in Nuuk, Greenland can be understood in Nome, Alaska. For millennia, the Inuit people from the Bering Strait to Greenland have largely lived peacefully.

That's enough history. Now a few personal photos of our short visit to Greenland.

This is the iconic view of Greenland: small, colorful houses hugging the seashore. It's still accurate for most of Greenland, but the country is beginning to change; a change that will only accelerate as interest in Greenland's strategic location and resources grows.

The view out of our hotel window of modern Nuuk towered over by Mount Sermitsiaq. Nuuk's traditional, colorful homes are being replaced by modern, but far less attractive apartment houses.

The pride of Nuuk, the Kaassassuk is Royal Greenland's latest, state-of-the-art fishing trawler. Royal Greenland is a $900 million company "that brings high-quality wild-caught seafood from the North Atlantic and Arctic Ocean to consumers around the world."

The "Cathedral of Greenland" was built in 1849 by Lutheran missionaries. "We have a lot of baptisms, we have a lot of confirmations, we have a lot of marriages. So, I'm not worried about the church," the church's pastor, Aviaja Rohmann Hansen, recently said.

The Lutheran church's sanctuary and altar.
Beautiful in its simplicity.

The first hole at the Nuuk Golf Club is a short par three. The fairways on the nine-hole course are bordered by rock outcroppings so accuracy matters. Hit a wayward shot and your ball bounces out of sight.

That's Nancy and me at Qooqqut Nuan, a summer camp along the Nuuk Fiord. We're wearing mosquito masks to protect ourselves from swarming mosquitos (actually small gnats). We were on a fishing trip up the fiord where Nancy caught a two-foot cod, seriously. Fishing in the fiord is unspoiled; drop a line and within seconds you've caught a fish. At the camp, a small restaurant cleaned and served the fish in a delicious meal topped off by dark, Norwegian beer.

# SHORT ESSAYS FOR INQUIRING MINDS

Even in mid-summer, our boat's captain had to be careful of icebergs quietly floating down the fiord from the glacier. This iceberg is small but extends 100 feet or more below the surface. Parts of the Nuuk Fiord are over 1,800 feet deep.

Ronald Gruner

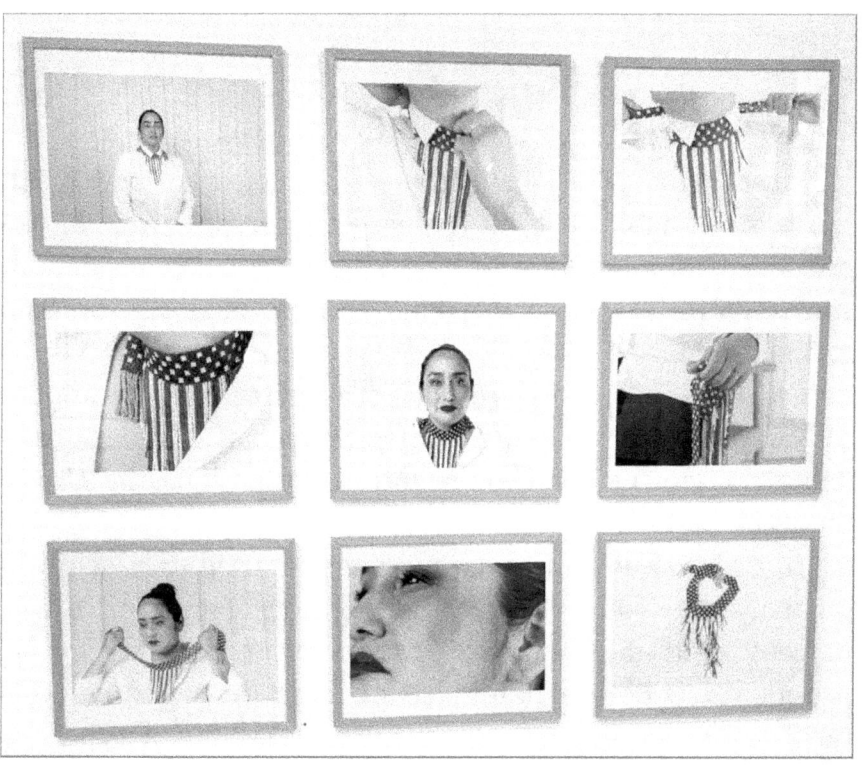

A collage of photographs in the Greenland National Gallery of Art. The artist has left its interpretation to the viewer.

# President Trump Announces New Tariffs and Fires Dr. Erika McEntarfer

AUGUST 5, 2025

Last Friday, President Trump announced his new tariffs will go into effect on August 7 rather than August 1 as previously announced. Tariff rates have evolved considerably since the President announced his "Reciprocal Tariffs" on April 2, Liberation Day.

Currently, a "Universal Tariff" of 10 percent will be applied to countries with which the United States has a trade surplus and a 15 percent tariff for countries with which the United States has a trade deficit.

In addition, President Trump has promised to raise tariffs on many other countries for various reasons. For example, a 50 percent tariff rate on Brazil for its prosecution of former president Jair Bolsonaro for election interference, a 39 percent tariff on Switzerland for its high U.S. trade surplus, a 35 percent rate on Canada for supporting a Palestinian state, and a 25 percent tariff on India for buying Russian oil.

Tariffs continue to be an evolving story. I'll write more about the tariffs as the situation stabilizes. For now, I'll just include an excerpt taken from my book, *We The Presidents*, discussing the 1930 Smoot-Hawley tariffs.

Most economists were against the [Smoot-Hawley] Act. As the legislation awaited Hoover's signature, 1,028 economists published an open letter urging the President not to sign the legislation. Henry Ford spent an evening with President Hoover urging him not to sign the bill calling it "an economic stupidity." Thomas W. Lamont, the chief executive of J. P. Morgan, warned the Act would intensify "nationalism all over the world." Undaunted, Hoover signed the Act into law on June 17, 1930, increasing average tariffs on over 20,000 imported goods from 41 to 47 percent. Other countries retaliated and raised their own tariffs. The impact was immediate. By 1932, American exports had fallen from $3.8 billion in 1930 to $1.6 billion, far deeper than the decline in GDP.

The other major event last Friday was the release of disappointing U.S. employment statistics by the U.S. Bureau of Labor Statistics (BLS). The BLS reported that during the first six months of the year, U.S. non-farm employment grew only 0.4 percent, from 158.9 to 159.5 million. During the same period in 2024, employment grew 0.7 percent—nearly double this year's rate.

In 1930, the Smoot-Hawley tariffs were controversial within the Republican party.

In the same report, the BLS announced that "employment in May and June combined is 258,000 lower than previously reported." The BLS attributed the revisions to "additional reports received from businesses and government agencies since the last published estimates and from the recalculation of seasonal factors."

President Trump was infuriated and abruptly fired Dr. Erika McEntarfer, the Commissioner of Labor Statistics at the BLS. "In my opinion, today's Jobs Numbers were RIGGED in order to make the Republicans, and ME, look bad," President Trump posted on Truth Social shortly after the BLS news release.

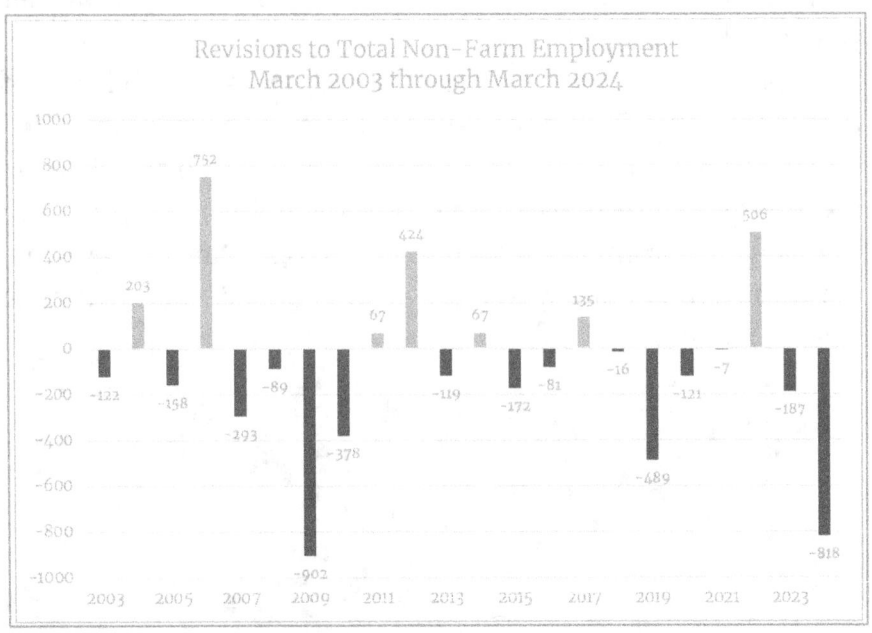

The Bureau of Labor Statistics often revises its initial employment statistics. BLS

But changes to BLS employment figures, both up and down, are common as the chart above shows. For example, in August 2024—three months before the presidential election—the BLS adjusted U.S. employment 818,000 downward suggesting the U.S. economy was not nearly as strong as President Biden had claimed. Similarly, early in President Obama's presidency, the BLS revised 2009 employment downward by 902,000.

It's important to understand that the BLS employment figures are estimates, and not actual employment numbers.

Every month, the BLS polls roughly 630,000 U.S. employers, from grocery stores to automobile manufacturers, to estimate the nation's non-farm employment. Unfortunately, the current system is imperfect: (1) like an election poll, the BLS only polls a fraction of all employers; (2) by law, employers, with a few exceptions, are not required to respond; and (3) response rates have declined from over 60 percent prior to the COVID pandemic to under 45 percent today.

Furthermore, employers are allowed, for any specific month, to submit two revisions. That's largely what happened last week when the BLS changed its employment figures for May and June after employers submitted revisions for those months.

Major revisions by the BLS tend to happen during periods of uncertainty—certainly the case today with President Trump's mercurial tariff policies. Similar to today, great uncertainty existed in 1980 and 1981 when the Federal Reserve raised federal rates to 19 percent to cool the economy in an effort to drive down inflation rates. Employers were uncertain, forcing the BLS to adjust its employment figures by over a million.

Other than President Trump, it's difficult to find anyone claiming the BLS is manipulating employment figures for political purposes.

*Fox Business* blamed the weak employment numbers not on politics, but on "elevated levels of economic uncertainty stemming from trade policy, including the impact of tariffs on inflation and consumer prices."

*The Wall Street Journal* summarized its view of President Trump's disparagement of the BLS last Friday when it declared:

Trump has had a long fixation on jobs numbers. During his 2016 campaign, he claimed that unemployment was far higher than official data showed. Once he took office, his view shifted. After the first full-month jobs report of Trump's first term, then-White House press secretary Sean Spicer said the jobs figures "may have been phony in the past, but it's very real now."

Even Steven Moore, one of the President's greatest supporters, blames the large BLS revisions on its methodology, not partisan politics. "A major reason [for the large revisions] is the response rate on the employer surveys is way down…. We hope the new BLS hire fixes the mechanics of job estimation."

Steven Moore is right. The process of estimating monthly employment is outdated and should be revamped. Why not, for example, use the monthly IRS tax filings as the basis for calculating true employment figures?

President Trump's abrupt firing of Dr. McEntarfer strikes me as exceedingly reckless. Unless his new BLS appointee confirms last Friday's employment statistics were either blatantly rigged or grossly incompetent, President Trump's abrupt firing of Dr. McEntarfer will be viewed as impulsive, petulant and mean-spirited—hardly presidential qualities.

# The Berlin Airlift

AUGUST 12, 2025

A friend of mine, Bibi LeBlanc, grew up in West Berlin during the 1960s and 1970s during the height of the Cold War. A wall ran through the center of the city. East of the wall lay communist Berlin, tightly controlled by the Soviets. To the west, free Berlin with a thriving, capitalistic economy.

Bibi had relatives in East Berlin she and her family would occasionally visit. East Berlin was a police state controlled by the *Stasi*, short for *Staatssicherheit*, the German word for State Security. Even crossing the border with its intimidating East German guards was stressful. (The Oscar-winning movie, The Lives of Others, depicts life in East Berlin under Stasi surveillance.) Bibi and her family were always happy, and grateful, to return home to West Berlin.

So it's not surprising that even as a child, Bibi knew she was lucky to be living in a free, democratic society. She also knew she owed her freedom to the Americans, and, in particular, one bold American operation shortly after the end of the Second World War: the Berlin Airlift.

In 1944 as the Second World War was coming to an end, President Roosevelt's Secretary of the Treasury, Henry Morgenthau, Jr., proposed a postwar plan to dismantle German industry "converting Germany into a country primarily agricultural and pastoral in its character." Unfortunately, the plan was widely publicized. Germany's Minister of Propaganda, Joseph Goebbels, capitalized on the publicity to spread anger and resolve within Germany by claiming the Allies intended to turn a defeated Germany "into a giant potato patch." One American colonel claimed Goebbels's threat was "worth thirty divisions to the Germans."

The deindustrialization of Germany remained American policy for the next three years. But by 1947, the growing Soviet domination of Eastern Europe had become a far larger threat

Uncle Wiggly Wings dropping candy to children during the Berlin Airlift.
BIBI LEBLANC

than a resurgent Germany.

President Truman's Secretary of State, George Marshall, proposed the United States reverse its German policy and rebuild Germany as part of a strong Europe. Congress agreed, and over the next four years, the Marshall Plan distributed

$12.7 billion ($170 billion in 2025 dollars) in financial aid to Europe.

But to rebuild Germany, more than financial aid was required. After the war, the German currency, the hyper-inflated Reichsmark, had become worthless forcing Germans to purchase goods through the black market, often with cigarettes. In early 1948, the United States and its allies replaced the Reichsmark with the Deutsche Mark.

Unwilling to adopt an American-backed currency, the Soviets objected. When the United States refused to withdraw the new currency, the Soviets closed all roads, railroads, and canals connecting West Germany with Berlin. Unless the United States was willing to start another war, the Soviets were confident they could starve the city, and the Americans, into submission.

But President Truman and General Curtis LeMay, head of the newly independent United States Air Force in Europe, had another idea: provide an air bridge from West Germany into Berlin to deliver food, medicine, coal, and other critical supplies. It seemed an outlandish idea. To sustain its 2.8 million inhabitants, West Berlin would need 6,000 tons of supplies delivered every day in all weather. This when the workhorse aircraft at the time, the Douglas C-47, could carry a mere three tons.

But America and its allies accomplished what few thought possible. Starting on June 24, 1948, hundreds and often over 1,000 cargo planes landed in Berlin every day delivering critical supplies. The Airlift fed West Berlin's 2.8 million citizens, kept West Berlin's economy functioning, and reinforced the broader Marshall Plan's promise of American commitment to rebuilding Europe.

But as much as a story of American resolve, Bibi's book is about American humanity. And there is no better example of that humanity than "Uncle Wiggly Wings" as Bibi describes him in her book.

> ...As he was leaving, Lt. Halvorsen reached into his pocket and pulled out two sticks of gum. The children divided the gum and even treasured the wrappers. Wanting to give them more, he promised to drop candy the next day, not at the fence but from the sky. Before leaving, a child asked how they would recognize him. With a smile, Halvorsen replied, "I'll wiggle my wings."
>
> Little did he know that this simple gesture would become a cherished symbol of hope during those dark days.
>
> ...[The next morning], as he approached Tempelhof in his C-54, he spotted the children eagerly waiting in the same spot. He rocked his wings, and his flight engineer dropped the small packages [of candy] attached to handkerchief parachutes.
>
> Word spread quickly, and more kids showed up every day. Soon, letters addressed to "Uncle Wiggly Wings" and "The Chocolate Pilot" started pouring in, but Halvorsen kept his candy drops a secret, fearing he might get into trouble. When his commander asked what he was doing, Halvorsen replied, "Flying, Sir..."
>
> Soon, many other pilots participated, and candy parachutes were assembled by school classes and church groups across the U.S. and sent to the children of Berlin.

*Wings of Freedom* can easily be read in an hour or two. It's fascinating and inspiring reading for anyone with an appreciation of history including middle-school students and above. It's especially a great gift for children and grandchildren.

As Denise Halvorsen Williams, the daughter of Lieutenant Halvorsen, wrote in the book's foreword, "As you immerse yourself in this story, I invite you to reflect on the enduring lessons of compassion, unity, and the indomitable human spirit. May the legacy of the Berlin Airlift inspire us all to strive for a better world, where kindness and generosity prevail."

After nearly a year, beaten and humiliated, the Soviets agreed to negotiate an end to the blockade on May 12, 1949.

The Berlin Airlift marked a turning point in the Cold War. The Soviet gamble failed, and West Berlin became a symbol of freedom deep inside the Iron Curtain. In the eyes of many Germans, the United States transformed from an occupying power into a trusted partner, a shift that shaped transatlantic relations for generations.

But beyond the Cold War, the Berlin Airlift was proof that logistics, aid, and good will could be as potent as military power. The Berlin Airlift influenced U.S. foreign policy for decades, from disaster relief missions to development programs that projected American influence through assistance rather than force.

The victory was achieved without firing a shot. American resolve and ingenuity had replaced bullets. In its resistance to a tyrannical adversary, in its logistical success, and in its magnanimity to a defeated enemy, the Berlin Airlift was one of America's finest moments.

Bibi LeBlanc tells this inspiring story in her new book, *Wings of Freedom*. Rather than a dry recounting of history, the book is written in a unique and compelling format. In over forty two-page spreads, each with a colorful illustration, the author describes, in both English and German, the politics, the people, the aircraft, the pilots, and the many other factors that contributed to the remarkable success of the Berlin Airlift.

# Is The Republican Pledge Obsolete?

AUGUST 19, 2025

For nearly forty years, Republicans seeking office have solemnly taken "The Pledge." In doing so, politicians promised never to raise taxes. pledging that:

> "I will: One, oppose any and all efforts to increase the marginal income tax rates for individuals and/or businesses; and Two, to oppose any net reduction or elimination of deductions and credits, unless matched dollar for dollar by further reducing tax rates."

Later versions of The Pledge simply declared, "I pledge… that I will oppose and vote against any and all efforts to increase taxes."

The Pledge originated during the Reagan presidency shortly after Grover Norquist founded Americans for Tax Reform (ATR) in 1985. Since its founding, ATR has advocated for "taxes [that] are simpler, flatter, more visible, and lower than they are today."

In Grover Norquist, President Reagan had found someone who distrusted government even more than he did. Beyond reducing taxes, Norquist wanted to drastically shrink government. "I don't want to abolish government," Norquist later declared. "I simply want to reduce it to the size where I can drag it into the bathroom and drown it in the bathtub."

Reagan's vice-president, George H.W. Bush, refused to sign The Pledge. "Who the hell is Grover Norquist, anyway?" Bush later asked. Bush had long been skeptical of Reagan's economic policies calling them "Voodoo Economics" during the 1980 presidential campaign.

Grover Norquist explaining his Taxpayer Protection Pledge in 2011.
NBC NEWS

Two years later, in 1988 during his own presidential campaign, Bush's refusal to sign The Pledge led to political attacks claiming that, as president, he would raise taxes. To his lasting regret, Bush then promised voters, "Read my lips, no new taxes." Bush broke that promise in 1990 when he signed the Omnibus Budget Reconciliation Act significantly raising taxes. Bush believed higher taxes were necessary to reduce the federal deficit which had ballooned during Reagan's presidency.

His broken promise cost Bush the 1992 presidential election.

After that, Republicans got the message. For decades, nearly every Republican running for office routinely signed Grover Norquist's Taxpayer Protection Pledge.

But support for The Pledge may be waning. Today, only 252 Republican senators, representatives, and governors have taken the pledge, just 84 percent of the 299 Republicans currently occupying those positions. The inexorable increase in the federal debt may be giving Republicans pause that, in addition to cutting expenses, taxes may need to be increased.

The tariffs that went into effect on August 1 may be the largest peacetime tax increase in U.S. history; economists are currently arguing that issue. But no person in Congress who signed The Pledge violated their oath. Ironically, after decades of taking The Pledge, Congress was never asked to vote on the new tariffs. President Trump, after declaring a national emergency, bypassed Congress and authorized the tariffs as simple executive orders.

No competent economist questions that tariffs are indeed taxes—taxes paid by the importer of goods purchased from foreign countries and then passed on to the consumer either directly or indirectly.

Assume, for example, that Walmart imports a very conservative estimate of $100 billion in foreign goods annually. Based on an average 18 percent tariff rate, Walmart would pay $18 billion in import tariffs for these goods.

Walmart can't "eat the cost" of the tariffs as President Trump has suggested. Why? During the last four years, Walmart's pre-tax income has averaged $21 billion. If Walmart were to absorb the $18 billion in tariff costs, its profits would be devastated destroying hundreds of billions in shareholder value.

If a giant like Walmart can't absorb their tariff costs, it's unlikely smaller firms—like your local grocery store chain—will be able to do so.

However, firms with few foreign imports (e.g. Tesla) or very high profit margins (e.g. Apple) may be able to absorb the tariff costs themselves. But even in these exceptional cases, doing so would reduce their profits needed to finance future expansion and provide a return to shareholders.

### U.S. International Trade in Goods: 2024
#### Sorted by Trade Balance

| | Imports | Exports | Trade Balance |
|---|---|---|---|
| China | 439,564 | 144,248 | -295,316 |
| Mexico | 515,587 | 334,436 | -181,151 |
| Vietnam | 136,518 | 13,043 | -123,475 |
| Ireland | 103,438 | 16,897 | -86,541 |
| Germany | 161,016 | 75,972 | -85,044 |
| Taiwan | 116,357 | 42,910 | -73,447 |
| Japan | 149,711 | 80,050 | -69,661 |
| Canada | 419,666 | 350,409 | -69,257 |
| Korea, South | 133,133 | 66,929 | -66,204 |
| India | 87,378 | 41,627 | -45,751 |
| Italy | 76,791 | 32,379 | -44,412 |
| Switzerland | 63,401 | 25,488 | -37,913 |
| Malaysia | 52,521 | 27,471 | -25,050 |
| France | 60,218 | 43,647 | -16,571 |
| Israel | 22,232 | 15,436 | -6,796 |
| All Other Countries | 757,650 | 768,834 | 11,184 |
| Total ( $ millions) | 3,295,181 | 2,079,776 | -1,215,405 |

The United States annually imports approximately $3.0 trillion in goods, all subject to potential tariffs. BUREAU OF ECONOMIC ANALYSIS

It's also unlikely that foreign exporters will be able, or willing, to lower their prices to mitigate American tariffs. World markets are highly competitive. If America's foreign suppliers could significantly lower their prices, they already would have.

So, with few exceptions, whether American companies pass the tariff costs directly to their customers or absorb the costs themselves, it's consumers and shareholders who ultimately pay for the tariffs.

Just how much money will tariffs raise? It's difficult to say, tariff rates remain in flux. In particular, tariffs for China are temporarily on hold at 30 percent through November 9, 2025. If by then a new agreement is not reached, President Trump has threatened to raise tariffs on Chinese goods to 145 percent. Similarly, Brazil and India have been threatened with 50 percent tariff rates, although many exemptions may apply.

What we do know is: (1) the Budget Lab at Yale and other respected sources estimate that tariffs on imported goods currently average about 18 percent, and (2) over the last five years the United States imported an annual average of $2.97 trillion in goods.

This suggests tariffs could raise as much as $530 billion annually in new federal revenues. That's a massive increase in federal revenues, an amount comparable to the total income taxes paid by the lower 85 percent of federal taxpayers in 2022 (the latest data currently available). It is $23 billion more than the $507 billion the IRS collected in corporate income taxes in 2024, and fifteen times the $35 billion collected in federal gasoline and diesel fuel taxes.

Of course, tariff revenue may actually be much lower. The original purpose of President Trump's tariffs was to encourage a rebirth of U.S. manufacturing. If that occurs, imports will decline along with tariff revenues. Tariff rates may also decline as the U.S. continues to negotiate with its trading partners. More foreign goods may be exempted. No one really knows.

But today, tariffs potentially are a massive windfall generating hundreds of billions in new federal revenues. How would Washington spend it? President Trump has suggested several possibilities including: (1) reduce or even eliminate income taxes for millions of lower and middle-income Americans, (2) provide direct rebates to citizens similar to the stimulus payments during the Covid pandemic, and (3) pay down the national debt currently at $37 trillion—$107,000 for every person in America.

Personally, I believe the best use for the new tariffs would be to help pay down our massive national debt. That's a tax even Grover Norquist should be able to support.

# "Gerrymandering is Incompatible with Democratic Principles."

AUGUST 26, 2025

Last week, the Texas State Legislature approved a mid-decade reorganization of five Congressional Districts currently held by Democrats. The reorganized districts are expected to strongly favor Republican candidates during the 2026 mid-term elections.

Democrats and much of the media have been indignant over last week's manipulation of electoral districts for political advantage. California Governor Newsom promised to retaliate with his own state's redistricting.

But partisan redistricting—practiced by both political parties—is hardly new, and dates back to the very founding of the United States.

Our nation's Founders wrote the Constitution with the intention that the House of Representatives would most closely represent the people through the direct election of Representatives, short terms, and small districts. Gouverneur Morris, one of the Constitution's principal framers, called the House "the grand depository of the democratic principle."

Those noble democratic principles were soon tested. During the first Congressional election in 1789 two American patriots, Patrick Henry and James Madison, were bitterly opposed in a fight over a Virginia Congressional seat.

Patrick Henry had been a Revolutionary War patriot famous for his declaration in 1775, "Give me liberty or give me death." But after the United States won its freedom from Great Britain, Henry feared America's new Constitution gave the federal government too much power. He believed that the Constitution "squints toward monarchy" and warned that "your president may easily become king."

Massachusetts Governor Gerry was immortalized when his redrawn state senate district was dubbed a "gerry-mander." WIKIPEDIA

James Madison was widely regarded as the author of the Constitution by the Founders. In contrast to Patrick Henry, Madison believed a strong federal government was necessary to keep the union from fracturing into rival states. Madison warned that without a strong federal government America could easily become a continent of small, competing states often at war—much as Europe had been for centuries.

The United States, now under its new Constitution, scheduled its first national election for February 2, 1789. James Madison ran as a candidate for Virginia's Fifth Congressional District in the House of Representatives. Patrick Henry was opposed to Madison's candidacy. He hoped the election would install candidates who, distrustful of the new Constitution, would call for a national convention to modify or even revoke it.

As the former governor of Virginia, Henry and his supporters had the political power to redraw the state's Fifth Congressional District to group Madison's home county with five counties whose voters generally opposed the Constitution. Patrick Henry believed that Madison could not win the election with voters from five of the district's six counties opposing him.

Patrick Henry's confidence was justified. Running against Madison was James Monroe, a respected Revolutionary War hero who had been wounded in the Battle of Trenton. Monroe shared the same distrust of the Constitution as Patrick Henry.

During his campaign, Monroe insisted that a Bill of Rights must be amended to the Constitution that guaranteed individul freedoms. Madison initially rejected the need for a Bill of Rights arguing that the Constitution was a charter of enumerated powers. Since the federal government was limited to what the Constitution explicitly allowed, "Why declare that things shall not be done which there is no power to do?" as Hamilton had written in Federalist Number 84.

But many voters didn't trust Madison's arcane legal argument and insisted on an explicit Bill of Rights. In an early example of American Democracy at work, Madison listened to voters and changed his position. He promised that, if elected, he would support a Bill of Rights.

Madison won the election by 336 votes (1,308 to 972).

After the election, Madison and Monroe quickly reconciled. Both men became the last Founders to serve as presidents, Madison from 1809 through 1817 and Monroe 1817 through 1825.

Although Patrick Henry's attempt to redraw electoral districts for political advantage failed, two decades later in 1812 Massachusetts Governor Elbridge Gerry tried again. That year, Governor Gerry drastically redrew the Massachusetts state senate districts hoping to tilt the election. The effort was only a partial success. Gerry lost his reelection but his party retained control of the legislature.

Governor Gerry's new senate district was so distorted that cartoons lampooned it as a mythological, winged salamander, or "gerry-mander." The name stuck, and ever since the practice of manipulating electoral boundaries has been known as gerrymandering.

For the next two-hundred years, politicians contorted their state and congressional districts to gain an electoral advantage over their political adversaries.

Before the Civil War, South Carolina and other Southern states drew districts to favor wealthy, slaveholding regions.

In the late 19th century, Republicans in Illinois created elongated, snake-like congressional districts that ran across the state east-to-west.

During the late Tammany Hall era (roughly 1900 through 1925), Democrats drew districts to concentrate Irish, Italian, and Jewish immigrant groups into Democratic strongholds.

But it wasn't until the last few decades, after computers simplified the design of complex political districts, that gerrymandering reached its pinnacle.

That pinnacle was arguably reached in 2011 when Pennsylvania Republicans so contorted the state's 7th Congressional District that it became known as the "Goofy Kicking Donald Duck" district. "Rather than keep communities intact and follow natural borders," *The New York Times* wrote, "the district slices and dices pieces of five counties and 26 municipalities and tosses them together in a crazy salad."

In January 2018, the Pennsylvania Supreme Court ruled the district was an unconstitutional partisan gerrymander and ordered that the district be redrawn on non-political lines.

Pennsylvania's gerrymandered 7th Congressional District reminded voters of Goofy kicking Donald Duck.

A year later, the Supreme Court accepted the case of Rucho v. Common Cause. The case was based on extreme partisan gerrymandering in North Carolina by Republicans and in Maryland by Democrats.

The Court acknowledged that partisan gerrymandering is "incompatible with democratic principles." But concluded that the Constitution does not provide a clear standard for determining when partisanship goes too far. In its 72-page decision, the Court commented:

> "The Framers [of the Constitution] chose a characteristic approach, assigning the [partisan gerrymandering] issue to the state legislatures, expressly checked and balanced by the Federal Congress, with no suggestion that the federal courts had a role to play."

Failing to rule on one of the worst abuses of the American political system, the Supreme Court kicked the issue of partisan gerrymandering back to the states. But partisan state legislatures ignored the Court's ruling and continued, even accelerated, to gerrymander their states. Our deeply divided Congress merely looked on.

### Examples of Gerrymandered States
#### 2024 Presidential Election

| State | Presidential Election percent | | House of Representatives percent (seats) | |
|---|---|---|---|---|
| | Democrat | Republican | Democrat | Republican |
| Massachusetts | 61% | 36% | 100% (9) | 0% (0) |
| California | 58% | 38% | 83% (43) | 17% (9) |
| Connecticut | 56% | 42% | 100% (5) | 0% (0) |
| New York | 56% | 43% | 73% (19) | 27% (7) |
| Illinois | 54% | 43% | 82% (14) | 18% (3) |
| North Carolina | 48% | 51% | 29% (4) | 71% (10) |
| Iowa | 43% | 56% | 0% (0) | 100% (4) |
| Texas | 42% | 56% | 34% (13) | 66% (25) |
| Indiana | 40% | 59% | 22% (2) | 78% (7) |
| Oklahoma | 32% | 66% | 0% (0) | 100% (5) |

U.S. HOUSE OF REPRESENTATIVES

Prior to last week's gerrymander vote by the Texas Legislature, Texas' representation in the House of Representatives was only moderately unbalanced. In 2024, Texas Democrats won 42 percent of the presidential vote but just 34 percent of the state's 38 House seats. Unfair? Yes, but consider California.

That same year, California Republicans won 38 percent of the presidential vote but only 17 percent of the state's 52 House seats. Other Democratic states were far worse, none more so than Massachusetts. There, Republicans won 36 percent of the presidential vote but none of the state's nine House seats.

So, it is not surprising that shortly after the Texas Legislature voted to redraw five Congressional districts along partisan lines, an elated President Trump posted on Truth Social:

> "Big WIN for the Great State of Texas!!! Everything Passed, on our way to FIVE more Congressional seats and saving your Rights, your Freedoms, and your Country, itself. Texas never lets us down. Florida, Indiana, and others are looking to do the same thing."

As the vigorous leader of the Republican party, Donald Trump's enthusiasm to further gerrymander Republican states, starting with Texas, is hardly a surprise.

But as our President, a better approach would have been to call for an end to all gerrymandering; let voters decide elections rather than political manipulation, manipulation "incompatible with democratic principles."

# Cracker Barrel Gives Uncle Herschel the Boot

SEPTEMBER 2, 2025

Last month was difficult, more like a nightmare, for Julie Masino, the CEO of the Cracker Barrel restaurant chain.

Starting in mid-2024, Cracker Barrel began to change its business model with updated menus and lighter and more modern store designs. Every business needs to evolve, and Cracker Barrel's changes appeared to be working. After releasing the company's third quarter financial results on June 5, CEO Julie Masino was optimistic:

> "Our third quarter performance exceeded our expectations and represents the fourth consecutive quarter of positive comparable store restaurant sales growth. We remain focused on executing our transformation plan and believe we are well-positioned to deliver a strong finish to the fiscal year."

Then, on August 19, Cracker Barrel announced it was changing its logo. Since 1977, Cracker Barrel's logo had featured Uncle Herschel sitting in a wicker chair leaning on a barrel presumably full of crackers. Uncle Herschel was founder Dan Evins' uncle on his mother's side. More than a marketing prop, Uncle Herschel helped shape Cracker Barrel's values during its early years according to the company's website.

The original logo conveyed a friendly country lifestyle. But the new logo was starkly generic: the restaurant's name in brown lettering against a yellow barrel shape. That's it. Uncle Herschel, his wicker chair and the old-time cracker barrel were gone.

Cracker Barrel's old and new logo.                    CRACKER BARREL

Within days, Julie Masino knew that she and her company had seriously misread the current political environment.

- In an X post, Donald Trump Jr. asked his ten million followers, "WTF is wrong with Cracker Barrel??!"
- After Cracker Barrel's stock crashed, political activist Robbie Starbuck posted, "Good morning, Cracker Barrel! You're about to learn that wokeness really doesn't pay."
- Suggesting Cracker Barrel's logo change had a political message, Fox News commentator Laura Ingraham asked, "...Was this actual hostility to traditional country culture?"
- A Hillsdale College social media post equated the new logo to desecrating a statue of George Washington with red paint.
- Replying to a post by President Trump, Sardar Biglari, an activist shareholder and CEO of Steak 'n Shake, posted, "...The woke CB management has even less competence than your predecessor. The CEO needs to hear 'You're Fired' from her board."

Millions of Americans were fed up with woke rebrandings. Land O Lakes butter dropped its image of a young Indian maiden. Uncle Ben's Rice dropped their kindly Black man renaming their product simply Ben's Rice. Aunt Jemima disappeared completely from the popular pancake mix, replaced by Pearl Milling Company. Cracker Barrel's logo change, millions believed, was just one more case of woke taking over corporate America.

On August 26, Cracker Barrel announced they were scrapping the new logo and returning to their original logo. "We thank our guests for sharing your voices and love for Cracker Barrel," the company posted. "We said we would listen, and we have. Our new logo is going away and our 'Old Timer' will remain."

After the announcement, the White House posted an image of President Trump replacing Uncle Herschel on the Cracker Barrel logo. "Go woke, go broke," the White House post proclaimed.

It's easy to criticize from the sidelines, but Cracker Barrel's logo designer made a serious mistake by not including Uncle Herschel in the new logo. Market research surely would have shown that customers identify with Uncle Herschel's friendly figure. Kentucky Fried Chicken has featured Colonel Sanders in its logo for over seventy years, as has McDonald's and its golden arches since 1961. Cracker Barrel's marketers must have known that abandoning that successful approach was risky.

With the exception of Steak 'n Shake's conflicted CEO, notably silent on the logo change were other company chief executives; most CEOs understood that a brand transition is one of the most difficult challenges any company can face. Many fail in the process.

Around 1970, the Boston Consulting Group (BCG) developed its Growth Share Matrix used to help large companies manage their various products, or in the case of a small business, themselves. The Matrix consists of four types of products (or single product companies like Cracker Barrel) based on their growth rate and market share.

- Stars: highly profitable products with high growth rates and high market share e.g. Apple iPhone in 2010.
- Cash Cows: profitable products with high market share but slowing growth e.g. Microsoft Windows in 2000.
- Question Marks: typically, new products with low market share but high growth rates e.g. Tesla in 2017.
- Dogs: typically older products with low market share and low growth e.g. Apple iPod in 2020.

Successful products, and companies, generally follow an arc starting from a Question Mark growing into a Star then declining to Cash Cow and ending as a Dog.

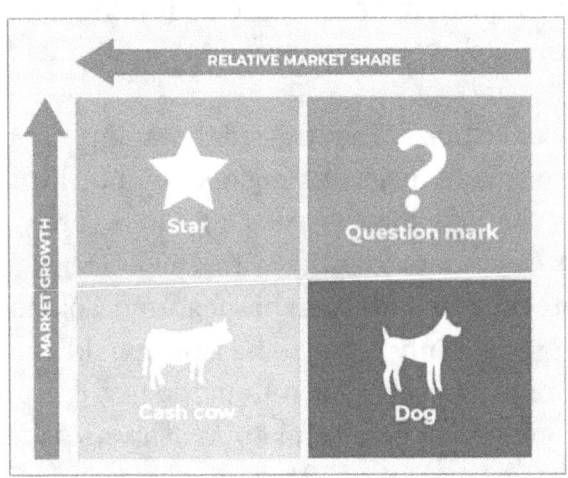

Boston Consulting Group's Growth Share Matrix

By this metric, Cracker Barrel is a Dog. Adjusted for inflation, Cracker Barrel's annual revenues and operating profits have declined for a decade.

Of course, Cracker Barrel's loyal customers don't see Cracker Barrel as a Dog. And it isn't for them with favorites such as Grandpa's Country Fried Breakfast, Biscuits n' Gravy, and its signature Southern Fried Chicken.

But focusing on a loyal customer base as markets evolve is dangerous. Companies with once large, loyal customer bases such as Blockbuster, Toys "R" Us, Kodak, and Borders have gone out of business as markets and customers change. The restaurant business is particularly fickle. Friendly's, Howard Johnson's, Steak and Ale, White Tower, Victoria Station and scores of other national chains, once beloved, are all gone. Only fast-food chains seem to survive more than a few decades: McDonalds, Dairy Queen, Kentucky Fried Chicken, Dunkin' Donuts...

So, Cracker Barrel, whose average customer is 54.7 years old, should be commended for attempting to adjust its business model to appeal to a broader and younger customer base.

What the company didn't anticipate was the huge backlash driven by the current anti-woke movement to eliminate Diversity, Equity, and Inclusion (DEI) policies in government, schools, and businesses.

But the warnings were there.

In 2023, after Bud Light beer used a transgender influencer, Dylan Mulvaney, to promote its brand on social media, conservative anger led to a nationwide boycott. Bud Light sales fell 25 percent costing the beer its twenty-year position as America's top-selling beer. Ironically, the American icon was replaced by Modelo Especial, a Mexican import brewed in Mexico City.

On August 10, just days before Cracker Barrel announced its logo change, Harley-Davidson fired its CEO after a social media firestorm erupted over its DEI policies. "Harley Davidson CEO Jochen Zeitz is officially OUT as CEO less than one year after I exposed Harley Davidson's woke policies..." prominent influencer, Robby Starbuck, posted on Facebook. "I warned that not firing CEO Jochen Zeitz meant doom for the company and I was right."

Disney, Harbor Freight, John Deere, Nike, Target and other companies have experienced similar social media pressure over their internal DEI policies.

Even before the current anti-woke campaigns, companies often stumbled making adjustments to their business model or products. In 1957, when Ford believed they needed to modernize their product line, the company introduced the Edsel. Customers hated the car's radical design with its "horse collar" grill. Ford dropped the Edsel three years later. Coca-Cola made an even worse mistake when it introduced a sweeter New Coke in 1985 and then, incredibly, dropped the original Coke. Customers revolted and "Coca-Cola Classic" was soon back on the market. Ford and Coca-Cola recovered from their mistakes, but one great brand did not, Schlitz beer. Once America's most popular beer, to increase production Schlitz changed the beer's formula in the mid-seventies. Customers could taste, and even see the difference. Sales collapsed and in 1982 Schlitz was forced to sell itself to Stroh Brewing Company.

Today's business environment is far more challenging than that navigated by Ford, Schlitz, and Coca-Cola. It is not just customers who revolt today over corporate missteps and policies, it's millions of outraged Americans consumed in the American Culture War.

"We boycotted Bud Light and made them lose billions for going woke," conservative commentator Rogan O'Handley (aka DC Draino) posted after Cracker Barrel's capitulation. "We boycotted Target and made them lose billions for going woke. Then we boycotted Cracker Barrel and they brought back the southern gentleman within a week. Maga is the most dominant political movement in our lifetimes. And we're just getting started."

# A Defiant RFK Jr., Poor Jobs Report, and New Department of War

SEPTEMBER 9, 2025

My last Substack focused on Cracker Barrel's troubled logo change. I hadn't eaten at one of their restaurants in decades, so last Wednesday had dinner at the Cracker Barrel restaurant in Tewksbury, Massachusetts.

I ordered country-fried steak (a southern version of Austrian wiener schnitzel) with white cream gravy, fried okra, and turnip greens along with a Budweiser beer. For a boy who grew up during the fifties in a small town in Oklahoma, it was delicious and inexpensive—a solid meal and cold beer for $20.19, not including tax and tip.

But the restaurant was nearly empty, too empty even for a Wednesday evening. And to me the decor, composed of memorabilia and nick-nacks hung on the walls, was tired and dated. So there's little question Cracker Barrel has some work to do, as do all restaurants who wish to survive with changing times.

I'll be going back for their authentic southern cooking and wish Cracker Barrel success in their corporate transition.

## Robert F. Kennedy Jr.

Last Thursday, Robert F. Kennedy Jr. testified before the Senate defending his leadership as the Director of Health and Human Services. It was a three-hour contentious meeting. During the meeting, Kennedy told senators:

- "The CDC failed that responsibility [to protect Americans] miserably during COVID," Kennedy said, adding that "oppressive and unscientific interventions failed to do anything about the disease itself."
- When asked how many Americans died of Covid, Kennedy responded, "I don't know how many died. I don't think anybody knows."
- When asked if he believed vaccines had prevented COVID deaths, Kennedy answered, "I would like to see the data and talk about the data."

| U.S. Health Outcomes vs Percent of Population with College Degree or More County Quintiles Ranked by Education Level | | | | |
|---|---|---|---|---|
| College Degree or More | Median Household Income (2019) | Life Expectancy (2019) | Heart Deaths per Million (2019) | Covid Deaths per Million (2020-2022) |
| 15% | 39,447 | 74.6 | 4,121 | 5,539 |
| 19% | 47,743 | 76.3 | 3,853 | 4,853 |
| 22% | 53,402 | 77.6 | 3,560 | 4,164 |
| 24% | 60,159 | 78.6 | 3,185 | 3,636 |
| 34% | 77,780 | 80.0 | 2,488 | 2,815 |

Deaths rates are, typically, lower for Americans with more education. CDC

In my book *COVID WARS*, I discuss Kennedy and his many theories extensively. During a May 2022 podcast interview, for example, Kennedy declared: "A global elite led by the CIA had been planning for years to use a pandemic to end democracy and impose totalitarian control on the entire world." And in the foreword for *The Truth about Covid-19*, Kennedy wrote: "The

COVID-19 vaccine may become the greatest public health disaster in history... You're going to see a lot of people dropping dead."

During the pandemic, Kennedy authored *Vax-UnVax: Let The Science Speak*; *The Real Anthony Fauci: Bill Gates, Big Pharma, and the Global War on Democracy*; and *Public Health, and the Terrifying Bioweapons Arms Race*. Throughout these books, Kennedy made incredible claims of evil conspiracies led by global elites, vaccines more dangerous than Covid itself, and that ivermectin and hydroxychloroquine cured and even prevented Covid.

Years later, none of Kennedy's conspiracy theories and grand claims have proven to be true.

The truth is simple. There were no global conspiracies, the vaccines were not deadlier than the virus, and Covid was not a hoax. Prior to his senate testimony, Kennedy could easily have generated the above table using his own highly trusted CDC data. (While writing *COVID WARS*, I often cross-checked CDC statistics with local health departments. There were very few differences since the CDC obtains its statistics from the nation's 3,300 local health departments.)

Rather than basing his testimony on crazy theories and condemnations of public health officials, Kennedy could have informed the senators with verifiable facts. The above table segments America's 3,142 counties into quintiles based on average education levels—obesity, smoking, vaccine hesitancy, and other health-related factors all decrease with education.

It's education, and affluence, more than any other factor which determines how we Americans manage our health habits, our ability to think critically in a environment flooded by disinformation, and ultimately our wellness and lifespan.

What American healthcare must do is to provide the same level of healthcare to all its citizens, not just those with the judgement and resources to live long, healthy lives.

## August Jobs Report

On Friday, the Bureau of Labor Statistics (BLS) issued its Employment Situation Summary, commonly known as its Jobs Report. It was disappointing.

"Total nonfarm payroll employment changed little in August (+22,000) and has shown little change since April..." the report declared. The poor report was a surprise. A Bloomberg survey of economists had estimated nonfarm payrolls would grow by 75,000 jobs in August. Instead, the economy added a meager 22,000 new jobs.

The BLS also revised its employment numbers for July and June. July employment increased from 73,000 to 79,000, but total employment in June decreased by 13,000 jobs. From April through August, the economy only added 107,000 new jobs—the lowest job growth since 2009 following the 2008 Financial Crisis.

But last Friday, President Trump did not blame the BLS for manipulating the jobs report as he did on August 1. That day, enraged over the low employment figures, the President abruptly fired the BLS Commissioner of Labor Statistics, Dr. Erika McEntarfer, a PhD economist who had spent over twenty years with the government. The President claimed McEntarfer had rigged the job numbers against him.

Rather than the BLS, President Trump accused Federal Reserve Chairman, Jerome Powell, for the poor numbers. Shortly after the report was released, President Trump posted: "Jerome 'Too Late' Powell should have lowered rates long ago. As usual, he's 'Too Late!'"

But as the expression goes, "Late's better than never." Friday's Job Report fueled expectations that the Federal Reserve would cut interest rates by a quarter percent or more in September, rate cuts President Trump has been advocating for months.

As I've written in previous Substacks, how Americans perceive major events such as Friday's disappointing Jobs Report depends on their news source. As of mid-morning Friday, the top headline from eight online news sites ranging from far right to far left (in my opinion) varied significantly based on their political orientation:

**Newsmax**
Trump: US Has 'Lost' India, Russia to China

**New York Post**
Top Trump adviser floats Eric Adams for Saudi Arabia ambassadorship to thin out busy NY mayoral field, give Cuomo a shot to beat Mamdani.

**Fox News**
Venezuelan military jets buzz US Navy ship in 'highly provocative' move, Pentagon says.

**Wall Street Journal**
Hiring Stalled in August, With 22,000 New Jobs.

**New York Times**
U.S. Labor Market Stalled This Summer, With August Data Adding to Slowdown.

**CNN**
America's job market flashes yet another warning sign about the economy

**MSNBC**
Trump's tariffs were supposed to revive U.S. manufacturing. They're wrecking it.

**HuffPost**
Trump Fired An Official Over A Weak Jobs Report. Another Brutal One Just Came Out.

The American media may have mixed views on the nation's economy, but consumers are in a very sour mood according to the University of Michigan's Consumer Sentiment Survey. The survey has been conducted since 1960 based on interviews that measure U.S. consumer attitudes towards personal finances, business conditions, and economic activity.

In the five months since President Trump announced his tariff policy on "Liberation Day," the survey has averaged 58.04 (1960=100). That is the most pessimistic level since early 1980 when inflation was 15 percent and unemployment was 7.5 percent.

## Department of War

Late Friday afternoon, President Trump announced he was changing the name of the Department of Defense to the Department of War. He stated his reason for the change in an executive order:

> "On August 7, 1789, 236 years ago, President George Washington signed into law a bill establishing the United States Department of War to oversee the operation and maintenance of military and naval affairs. It was under this name that the Department of War, along with the later formed Department of the Navy, won the War of 1812, World War I, and World War II, inspiring awe and confidence in our Nation's military, and ensuring freedom and prosperity for all Americans. The Founders chose this name to signal our strength and resolve to the world. The name "Department of War," more than the current "Department of Defense," ensures peace through strength, as it demonstrates our ability and willingness to fight and win wars on behalf of our Nation at a moment's notice, not just to defend."

Branding might sell condominiums; I doubt it will help win wars. But it might lead to hubris, a fault long too common in American military thinking.

At the beginning of the Vietnam War, when asked how long the war might last, Ronald Reagan responded: "It's silly talking about how many years we will have to spend in the jungles of Vietnam when we could pave the whole country and put parking stripes on it and still be home for Christmas." Four decades later as America was going to war in Iraq, Secretary of Defense Donald Rumsfeld promised Americans, "I can't tell you if the use of force in Iraq today will last five days, five weeks or five months, but it won't last any longer than that."

Reagan, Rumsfeld, and many others believed America could "fight and win" these wars in months. After many years and over 60,000 American deaths, both ended in stalemate, at best.

# Charlie Kirk and Martin Luther King Jr.

SEPTEMBER 16, 2025

The murder of Charlie Kirk last week brought memories back of another political assassination almost sixty years ago, Martin Luther King Jr. who was killed on April 4, 1968. Both men led transformative national movements: King in civil rights and Kirk in Christian nationalism.

I first heard Martin Luther King Jr. had been murdered when I arrived at work the next day. I was working at a General Electric factory in Oklahoma City while half-heartedly attending college. Somehow, I missed hearing the news. With no internet, news travelled slower in those days.

I remember my work buddies and me sharing thoughts about King, but mostly about what was happening in America. All of us, of course, remembered President Kennedy's assassination on November 22, 1963. I was in eleventh grade English class when the news was announced on the PA system. Class was adjourned and we were directed to the school chapel to pray for the president.

President Kennedy's death was a national tragedy that nearly all Americans, regardless of political persuasion, grieved over. In a *New York Daily News* article, Barry Goldwater, the political antithesis of John Kennedy, said, "The president's death is profound loss to the nation and the free world. He and I were personal friends. He is also a great loss to me."

Today, it is hard to understand how shocking Kennedy's assassination was to most Americans. It had been a century since Abraham Lincoln was assassinated near the end of the Civil War. (We had forgotten, or never knew, that Presidents James Garfield and William McKinley had also been assassinated.) In 1963, the assassination of an American president was a nearly unimaginable event.

Kennedy's death was just the beginning to a decade of violence.

Martin Luther King Jr. and Charlie Kirk
MARTIN OCHS ARCHIVES | TURNING POINT USA

On July 2, 1964, President Johnson signed the Civil Rights Act that outlawed discrimination based on race, color, religion, sex, or national origin. The Act deeply divided America.

"It is an act of tyranny," Alabama Governor George Wallace declared. "It is the assassin's knife stuck in the back of liberty.... It threatens our freedom of speech, of assembly, or association, and makes the exercise of these Freedoms a federal crime under certain conditions." Restaurant owner and future Georgia governor Lester Maddox famously chased out three Black men waving a pistol. "This property," Maddox said, "belongs to me and I'll throw out a white one, a black one, a red-headed one or a bald headed one. It doesn't make any difference to me."

Malcolm X was murdered on February 21, 1965. Born Malcolm Little, Little was a controversial Black leader who championed Black empowerment, pride, and resistance against racial injustice. Despised by social conservatives, Malcolm X shrewdly adopted Barry Goldwater's controversial declaration, "Extremism in the defense of liberty is no vice; moderation in the pursuit of justice is no virtue," as his own.

The murder of Malcolm X was the beginning of a bloodbath against civil rights advocates. On February 26, 1965, Jimmie Lee Jackson was shot by state troopers while attempting to protect his mother and grandfather during a civil rights march. On March 11, the Reverend James Reeb was beaten to death while participating in the Selma Freedom Marches. On March 25, Viola Liuzzo was shot and killed by a Klansman in a passing car while she was ferrying Selma Marchers between Selma and Montgomery. On June 5, Oneal Moore, a newly hired Black deputy, was killed by a shotgun blast from a passing car. On July 18, Willie Brewster was killed coming home from work by members of the National States Rights Party. On August 20, Jonathan Daniels, an Episcopalian seminary student helping with Black voter registration, was killed by a deputy sheriff. On January 3, 1966, Samuel Younge, Jr. was killed by a gas station attendant when he objected to segregated restrooms. On January 10, Vernon Dahmer was burned to death after his home was firebombed the day after a radio station broadcast his promise to pay voter poll taxes for those unable to afford them.

The murders continued through 1967 and 1968, culminating in the assassination of Martin Luther King Jr. King was a follower of Mahatma Gandhi who, after decades of nonviolent resistance and peaceful protest, wrested India from the British. Like Gandhi, King encouraged patient, non-violent protest against racism. In 1964, King won the Nobel Peace Prize.

America's political leaders reacted to King's death with restraint urging calm and even forgiveness. Just hours after

King's murder, Robert Kennedy—then running for president—gave a moving speech in Indianapolis declaring:

> "What we need in the United States is not division; what we need in the United States is not hatred; what we need in the United States is not violence or lawlessness, but is love and wisdom, and compassion toward one another, and a feeling of justice towards those who still suffer within our country, whether they be white or whether they be black."

Richard Nixon, who was running against Robert Kennedy for the presidency, visited the King family two days after King's death to offer his condolences and attended the funeral. Although Nixon had often been critical of the civil rights movement, he chose not to politicize King's death.

President Johnson declared April 7 a National Day of Mourning and ordered flags to be flown at half-staff on federal buildings. In a short, somber statement, Johnson urged:

I know that every American of good will joins me in mourning the death of this outstanding leader and in praying for peace and understanding throughout this land. We can achieve nothing by lawlessness and divisiveness among the American people. It is only by joining together and only by working together that we can continue to move toward equality and fulfillment for all of our people.

Two months later on June 5, 1968, Robert Kennedy was murdered in a Los Angeles hotel kitchen after addressing his campaign supporters. Like Charlie Kirk's death, King's death broke the hearts of millions of young Americans.

Today, America has been embroiled in nearly two decades of both political and seemingly random violence.

In 2009, thirteen were killed during the Fort Hood Shooting, and another thirteen killed during the Binghamton Immigration Center Shooting. In 2011 at a Tucson, Arizona political event, six were killed and Representative Gabby

Giffords is seriously wounded. In 2012, twenty children and six adults were killed in the Sandy Hook Elementary School shooting. In 2016, forty-nine were killed in the Pulse nightclub shooting. In 2015, nine Black parishioners were killed during the Charleston church shooting. In 2017, sixty were killed in Las Vegas, the deadliest mass shooting in modern U.S. history. Also in 2017, Representative Steve Scalise was critically wounded during a shooting at baseball practice. In 2018, eleven were killed in the Tree of Life Synagogue shooting. In 2022, nineteen children and two adults were killed in the Robb Elementary School shooting. In 2024, President Trump was nearly killed in an assassination attempt and a second plot was foiled by law enforcement. In 2025, former Minnesota House Speaker Melissa Hortman and her husband were killed in their home; three months later Charlie Kirk was killed at Utah Valley University.

Charlie Kirk's murder was a tragedy, a tragedy for his wife and two children, and for the millions who loved and supported him. But equally tragic, were the murders of those who have died in schools, churches, while working, at home, or simply being in the wrong place at the wrong time.

After Charlie Kirk's death, the possibility that we Americans will have a national dialogue on why we are so anxious to kill each other—whether school children or presidents—seems even more remote.

I won't quote the accusatory comments many of our political leaders have made since Kirk's murder. Most subscribers, I'm sure, have already seen them. Politics divides us, but cable and social media are its megaphones.

# Charlie Kirk in His Own Words

SEPTEMBER 23, 2025

Charlie Kirk began his transformation into a conservative populist when he was thirteen years old. Here is how Kirk described his political awakening in an April 29, 2025, podcast with Dr. Jordan Peterson, a Canadian psychologist and social media commentator:

> "I remember the discussions we'd have in class were, were very much anti-colonialist, anti-Western. That we [western civilization] are the contaminants on the world. That we are polluting other tribes. And again, right, mind you that we're in 8th or 9th grade discussing this. So, I don't know what post structuralism is or postmodernism is, but it was pre-woke. We weren't quite there though. And so, from the historical standpoint it was not [the] 1619 project, but it was. We're going to spend a whole month on slavery and we're going to spend three days on the founding."

Five years later, after spending one semester at Harper College, Kirk dropped out to become a full-time political activist.

The beginning was modest. As Kirk later said, he had "no money, no connections, and no idea what I was doing." Kirk would set up a card table at a nearby college, post a sign such as "I think government should be smaller," and offer to debate anyone. At first, only a handful of curious students might stop to talk. But Kirk was engaging and well-informed. Soon small crowds were waiting to debate Kirk on nearly any topic.

Kirk's growing popularity caught the attention of William Montgomery, a retired businessman and conservative activist. Montgomery encouraged Kirk to become a full-time political activist rather than return to college. In 2012, Kirk and Montgomery co-founded Turning Point USA (TPUSA) with a mission to advocate conservative values on high school, college, and university campuses.

Charlie Kirk speaking at the 2020 Republican National Convention
PBS NEWSHOUR

In 2016, Kirk voted for Donald Trump although he admitted he was "not the world's biggest Donald Trump fan." Kirk viewed Donald Trump as a modern-day Biblical Samson, a flawed but defiant man chosen by God to destroy the Philistines in a self-sacrificing act of destruction. In Trump's case, the Philistines were the Deep State.

In 2018, Kirk began to broadcast his debates on social media greatly expanding his reach, and financial donations. According to its 501c non-profit filings, TPUSA's revenues ballooned from $8.3 million in 2017 to $39.8 million in 2020.

A rising young conservative star, Kirk spoke at the 2020 Republican Convention. It was a break-out performance in which Kirk strongly endorsed Donald Trump. Referring to Trump's June 2015 descent down the golden escalator to announce his presidential candidacy, Kirk declared:

> "We may not have realized it at the time, but Trump is the bodyguard of western civilization. Trump was elected to protect our families from the vengeful mob which seeks to destroy our way of life, our neighborhoods, schools, churches, and values. President Trump was elected to defend the American way of life.... This election is not just the most important of our lifetime, it is the most important since the preservation of the Republic in 1865."

Now merged with Trump's MAGA agenda, TPUSA grew rapidly producing documentaries, hosting conferences, and organizing campus chapters across the country. By 2024, TPUSA was reporting $85 million in revenues. In its 501c filing, TPUSA claimed 1,873 campus chapters devoted to promoting conservative values. The filing noted that TPUSA's Faith division "is leading a movement to push back against secular totalitarianism in America, eradicate wokeism from the church, inspire the rise of strong churches, and wake up believers to their biblical responsibility to fight for freedom."

Over his short twelve-year career, Charlie Kirk's politics evolved from the Tea Party movement promoting personal liberty and smaller government to Donald Trump's populist MAGA ideology and, finally, to Christian nationalism—that the United States is divinely favored and should be governed by Christian principles.

Kirk believed the Founders envisioned a Christian form of government rather than the separation of church and state. Kirk makes this claim, for example, in an August 6, 2024, 'X' video post which asserted, "America was founded as a Christian nation. Prove me wrong."

The video, posted by Charlie Kirk himself, is an example of how Kirk often debated with his opposition. In the video, Kirk begins with an avalanche of assertions and ends with the declaration:

> "One of the reasons we're living through a constitutional crisis is that we no longer have a Christian nation, but we have a Christian form of government, and they're incompatible. You cannot have liberty if you do not have a Christian population."

Was Charlie Kirk right? Should Americans actually have a Christian form of government?

In 1785, James Madison, known as the father of the Constitution, responding to a Virginia legislature bill to fund Christian teachers, argued that government support for religion corrupts both government and faith. So not surprisingly, the Constitution never mentions God. The Constitution's Article VI even states "no religious Test shall ever be required as a Qualification to any Office or public Trust under the United States." In addition, the First Amendment reenforces the government's secular role: "Congress shall make no law respecting an establishment of religion or prohibiting the free exercise thereof." In 1802, President Jefferson famously confirmed the government's secular role. His "letter to the Danbury Baptists," described "a wall of separation between Church & State" assuring the congregation of the government's non-interference in their religious practices.

So clearly, although the Founders were Christians, they deliberately chose to separate church and state in America's new government.

Yet, Charlie Kirk asserted that the church and state are inseparable.

Beyond that, Kirk supported the "Seven Mountains Mandate," a theology that emerged in the nineteen-seventies. The Mandate requires that Christians "take dominion" over the seven "mountains" of society: government, education, media, entertainment, business, family, and religion.

The Mandate's adherents believe its justification is found in Biblical texts such as Isiah 2:2, "...let us go up to the mountain of the Lord, to the house of the God of Jacob; and he will teach us of his ways, and we will walk in his paths."

According to the Mandate, Christian theology should form the foundation for all of American society—an interpretation similar to Islam's Sharia law which specifies that Muslims live according to God's will as defined in the Quran.

Charlie Kirk's religious beliefs shaped his views on many issues. Below is a representative sampling—I believe—of Kirk's comments taken from recent videos.

## Abortion

"We allow the massacre of a million and a half babies a year under the guise of women's reproductive health. We are allowing babies to be taken away and discarded every single year, just saying they are not humans.... That's how we got Auschwitz, that's how we got the greatest horror of the 20th century."

## LGBTQ (Lesbian, Gay, Bisexual, Transgender, or Queer)

"I want you to be very cautious putting drugs into your system in the pursuit of changing your body. Instead, I encourage you to work on what's going on in your brain first. I think what you need first and foremost is just a diagnosis, just someone that is going to listen to what you have gone through, listen to what else is going on. My prayer for you, and again very few will say this, I actually want to see you be comfortable in how you were born."

### Diversity, Equity, and Inclusion
"As an organization you must prioritize something. We think that excellence should be an uncompromising pursuit and commitment to excellence and not equity. And we believe tht the more we embrace the mediocre, this kind of culture mediocrity, excellence will slip.... We don't believe in racial quotas. We want excellence quotas..."

### Family and Women's Role
"Shouldn't women stay at home, and have children, and do what they are designed to do?.... Wouldn't that make them happier?.... Maybe men are upset because the women that they are trying to date are more interested in taking care of cats and trying to become partner at the local law firm and they say I don't want to get married until I'm thirty and that creates a sense of despondency when a young male raised in this country sees everything rigged against him."

### Gun Control
"The only reason you want to register the guns is if you want to confiscate them.... I tend to be much more on the Libertarian side, although I'm not Libertarian, we actually need looser gun laws."

Earlier this year, Charlie Kirk proudly released "The Two Billion View Video." The video is a compilation of snippets from Kirk's debates with students during 2024. If you want to understand Charlie Kirk, watch his short, twenty-five-minute video. (After Kirk's death, the *New York Post* released its own video of Charlie Kirk's "Best Debate Moments.")

Charlie Kirk's September 21 memorial service attracted an estimated 100,000 mourners. During the five-hour service, Erika Kirk said her husband wanted to save lost young men, like the man accused of taking Charlie's life. "That man, I forgive him," she said quoting Jesus who asked God to forgive those who crucified him "for they know not what they do."

# The Boston Crusaders

SEPTEMBER 30, 2025

Boston has always been an avid sports town blessed with great sports teams: the Red Sox, Patriots, Celtics, and Bruins. In the last fifty years, the Red Sox won four World Series, the Patriots six Super Bowls, the Celtics seven NBA Championships, and the Bruins three Stanley Cups.

Now after fifty-three years of competition, another Boston team has won their first national championship, the Boston Crusaders.

Never heard of the Crusaders? I hadn't either until a good friend, Kevin Russell, introduced me to the national drum and bugle corps competitions. Yes, there is such a competition. America's young drum and bugle corps are a remarkable American subculture that has blossomed every summer since the formation of Drum Corps International (DCI) in 1972. For over fifty years, DCI's mission has been to promote youth-focused competitive drum corps events throughout the United States.

Like cheering for the Chicago Cubs, Kevin has been an ardent Boston Crusaders fan for years. Each summer, Kevin would spend weeks on the road travelling across the country to follow the Crusaders as they competed for Drum Corps International's grand prize.

But never won, until August 9, 2025.

The road to the Crusaders' victory began on May 23rd when the 165 members of this year's corps assembled at Vermont

State University in Castleton, Vermont to begin five weeks of intensive training on their twelve-minute routine. Each member, whose age is limited to twenty-one and younger, had been selected after an intense audition and interview process.

After training and then participating in several local events as dress rehearsals, the Crusaders began their national tour on June 28th when they competed in Fort Collins, Colorado. Over the next six weeks, the Crusaders competed in twenty competitions travelling by bus across the nation from Pasadena, California to Lawrence, Massachusetts. The final two-day competition was in Indianapolis where twelve teams competed for the gold medal in DCI's World Class Finals. After fifty-three years, the Boston Crusaders finally won, edging out the Canton, Ohio Bluecoats.

The Boston Crusaders' roots trace back to early colonial drum and fife corps. BOSTON CRUSADERS

But unlike the great sports teams, there was no prize money, no national endorsements, no trips to the White House. After a celebratory banquet the next day, the Boston Crusaders and its staff members quietly returned home.

It seems appropriate that a drum and bugle corps from Boston would finally be titled the best in America since it all started in Colonial Boston. In 1775 during the British siege of Boston, General Washington ordered each regiment in the

Continental Army to maintain a drum and fife corps. During marches, the drummers and fifers would set the cadence. During battle, they would provide signals to advance, charge, and when necessary, retreat.

By the early nineteenth century, starting with the War of 1812, the bugle had largely replaced fifes. A bugle cuts through the chaos of battle far better than a fife. During the Civil War, bugle calls such as reveille, taps, tattoo (return to barracks), charge, and retreat had become standardized, and played an important role in battle. (What young boy growing up in the fifties—Native Americans notably excepted—wasn't inspired by the movies depicting the U.S. Cavalry galloping to the rescue of a besieged wagon train with bugles blaring and flags flying?)

With thousands of soldiers trained on the bugle, brass bands naturally followed. John Phillip Sousa, the "American March King," led the charge. From 1880 through 1892, Sousa directed the U.S. Marine Band transforming it into the nation's premier military band which it remains to this day.

After resigning his Marine commission, Sousa formed his private Sousa Band. For the next forty years, until his death in 1932, the band toured the world giving over 15,000 band concerts. A prolific composer, Sousa wrote 136 military marches including the popular Stars and Stripes Forever, Semper Fidelis, and Washington Post March.

By 1932 when Sousa died, thousands of high schools and colleges had their own bands giving concerts, marching in parades, and performing during football games.

Eclipsed by the more colorful and entertaining brass bands, civilian drum and bugle corps largely disappeared. They had already been displaced in the military when telephone and radio signals replaced bugle calls to coordinate battle operations.

But in 1940, a new drum and bugle corps, the "Most Precious Blood Crusaders"—associated with the Most Precious Blood parish in Hyde Park, Boston—began a revival. By the late nineteen-forties, the corps, now independent, was known simply as the Boston Crusaders. The new Crusaders were, arguably, the first independent civilian drum corps not associated with a church, school, or veterans' post.

Over the next two decades, the Crusaders merged the colorful routines of the new marching bands with the regime of traditional drum and bugle corps to create their own unique musical and visual style; a style that pioneered the pageantry of today's drum corps competitions. Those competitions began in August 1972 when the newly formed Drum Corps International sponsored the first drum corps competition in Whitewater, Wisconsin.

For the next fifty-three years, the Boston Crusaders dutifully competed in the national competitions, but never won—until August 9, 2025.

They say the best form of leadership is by example. For me, the Boston Crusaders and the forty-seven other teams that competed in the 2025 DCI championships represent the best of America. They were chosen on merit across all races and ethnicities, trained intensively, had no political agenda, evangelized no religion, received no compensation, were gracious in defeat and victory, and when the competition was over, like colonial soldiers, quietly returned home.

# Congress Shuts Down the Government

OCTOBER 7, 2025

Last Wednesday, October 1, the federal government shutdown. Neither Democrats nor Republicans were able to reach sixty votes in the Senate to pass a federal funding bill—much less a bipartisan bill. Approximately 800,000 federal employees have been furloughed, sent home with no pay. Employees working in critical positions, such as air traffic control, must remain at work, but with their pay suspended until the shutdown ends.

Members of Congress (but not their staff) will continue to be paid throughout the shutdown. In 1983, Congress passed a permanent appropriation funding Congressional salaries in perpetuity.

From 1789 through 1979, the federal government never shutdown. In the rare instances when Congress failed to pass a funding bill, federal agencies minimized non-essential services but remained open, paying workers knowing Congress would soon appropriate funding.

But starting in the nineteen-seventies, Congress began to hold funding bills hostage as it grappled with contentious issues such as abortion, deregulation, and human rights. By 1980, these "funding gaps" had become so common that President Carter asked Attorney General Benjamin Civiletti to issue an opinion regarding whether spending money that Congress had not appropriated violated the 1884 Anti-Deficiency Act. Civiletti quickly concluded that, indeed, spending money not yet appropriated by Congress was illegal.

That was the birth of the government shutdown.

Shortly after Civiletti issued his opinion, the government shutdown on May 1, 1980, for a single day. The idea of the United States government actually shutting down was so radical, so irresponsible, so unpatriotic, that an ashamed Congress quickly appropriated funding.

Politics is a cynical business.                    U.S. CONGRESS

Since that first, short shutdown, the U.S. government has shutdown ten times for a total of ninety-three days (through October 6). Forty-four days have been during President Trump's two presidencies including the thirty-five-day shutdown, beginning on December 22, 2018, over funding for the Mexican border wall.

The current shutdown is over a dispute regarding expiring tax credits for health insurance and funding cuts to Medicaid contained in President Trump's One Big Beautiful Bill. Democrats want to retain the tax credits and reverse the cuts to Medicaid. Republicans disagree, insistent on reducing federal spending.

Republican efforts to reduce the federal deficit (the amount spending exceeds tax collections) by cutting spending is understandable. Since 2001, the annual deficit has increased from 1.0 percent of the Gross Domestic Product (GDP) to over 6.0 percent today. In 2024, for example, this flagrant overspending required the federal government to borrow $1.8 trillion to pay its bills—$5,300 in new federal debt added for every man, woman, and child in the United States.

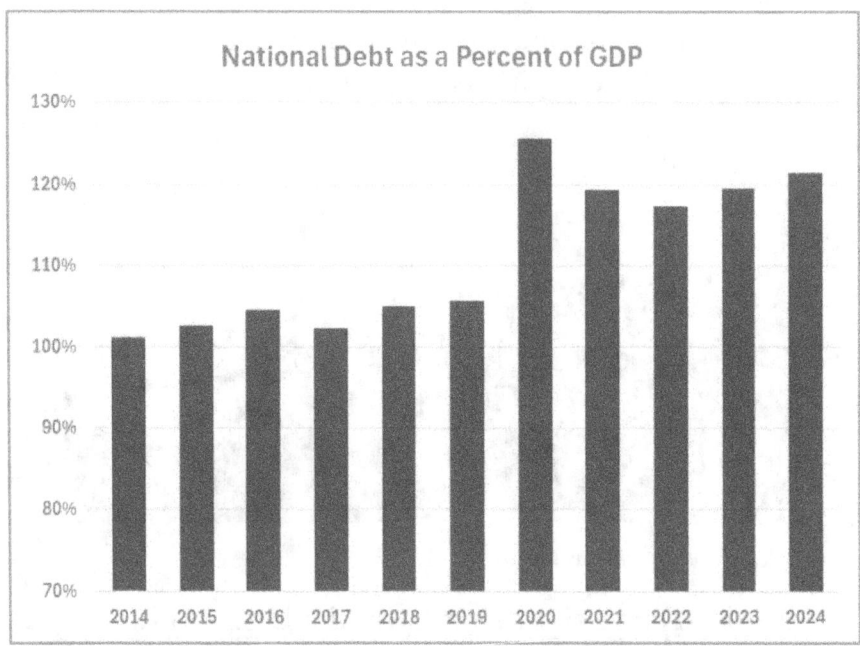

The U.S. national debt as a percentage of GDP is the highest since the end of the Second World War.     U.S. DEPARTMENT OF THE TREASURY

The Bipartisan Policy Center projects the 2025 deficit to be over $2.0 trillion. Yet, even with deficits skyrocketing President Trump's One Big Beautiful Bill signed on July 4th has been promoted as "the largest tax cut in history" while authorizing billions in additional spending from "creating Trump Accounts for every American newborn" to "funding the Golden Dome missile defense system to confront 21st century threats."

The bill does promise to "Restor[e] fiscal sanity by cutting $1.5 trillion in spending" although where and over what period is not specified in the White House news release.

One of the most successful federal budget interventions in the last century was President Reagan's management of the Social Security crisis; a problem Reagan inherited from President Carter who had attempted to fix the problem but failed.

As Reagan took office, Social Security was headed for bankruptcy. Having promised never to raise taxes, Reagan initially proposed cutting benefits as much as 30 percent. Millions of older Americans were enraged. Reagan quickly pivoted. As I wrote in, *We The Presidents*:

> "To broaden Social Security's tax base, [President Reagan's] commission proposed extending Social Security to newly hired civilian federal employees and employees of non-profit associations, the reduction of "windfall" benefits for those who contributed little in Social Security taxes, a delay in cost-of-living increases, an increase in the combined payroll tax to 12.4 percent, taxing a portion of Social Security benefits, reducing benefits and gradually increasing the retirement age.
>
> "These were difficult changes, but the sacrifices were, as equitably as possible, spread across workers, employers, and beneficiaries....
>
> "Social Security quickly returned to solvency. As Reagan was leaving office in 1989, Social Security was generating a $50 billion annual surplus and climbing."

That's real leadership. Leadership that is sorely needed today. Today's federal deficit is so large, no single fix such as cutting "waste, fraud, and abuse" will correct the problem.

Even if the "Department of Government Efficiency" (DOGE) somehow fired half of the 2.4 million federal civilian workforce, the savings would hardly impact the deficit. In 2024, the average compensation for civilian federal employees, including salary and benefits, was approximately $110,000 according to the Congressional Budget Office. So, firing half the federal civilian workforce—1.2 million employees—would only save $132 billion, about 2 percent of the $6.8 trillion in federal expenditures during 2024.

So, President Trump, like President Reagan, must not only reduce spending, but almost certainly increase taxes and find new sources of revenue.

Just how much would the federal deficit need to be reduced to bring the federal budget into balance? The perfect solution would balance the budget; federal revenues would equal, or even exceed, expenditures. From 1998 through 2001, President Clinton, working with House Speaker Newt Gingrich, successfully balanced the budget for the first time since the Coolidge administration during the Roaring Twenties.

But Clinton and Gingrich had a strong tailwind as I wrote in, *We The Presidents*:

> "The Cold War had ended, allowing reductions in defense spending. Baby Boomers, in their peak earning years, were generating large Social Security tax surpluses. A technology boom drove employment, corporate profits, and the stock market to record highs."

Other than a technology boom driven by Artificial Intelligence, those favorable conditions don't exist today. Today, defense spending is climbing and Baby Boomers are retiring in massive numbers.

But Congress doesn't need to balance the budget to restore fiscal responsibility. It simply needs to slow the growth in federal expenditures to equal or, ideally, be smaller than the growth in the nation's Gross Domestic Product.

A family's credit card balance can safely increase 3 percent a year if their family income is increasing 6 percent annually. The same logic applies to the government.

From 2014 through 2024, the GDP grew annually at a 5.7 percent average rate—a healthy growth rate even after inflation. But the national debt grew significantly faster, an average of 7.8

percent annually. During those years, federal expenditures consistently exceeded federal revenues from taxes and other sources.

By 2024, the federal debt, measured as a percentage of GDP, had increased from 101 percent in 2014 to 121 percent in 2024. (Up from 55 percent in 2000 before the Middle East Wars and 2008 Financial Crisis.)

For decades, Congress has financed government spending by borrowing billions and now trillions of dollars. Buyers of U.S. debt trust interest payments will be met and the principal will be repaid.

It is that trust that keeps the federal government running.

But trust is easily lost, and when it is, that loss happens quickly as it did during the 2008 Financial Crisis, the 1987 Savings and Loan Crisis, the 1928 Stock Market Crash and Great Depression, and the Financial Panic of 1907. As Hemingway wrote in *The Sun Also Rises*: "How did you go bankrupt? ...Two ways. Gradually, then suddenly."

# Just How Serious is the Federal Deficit?

OCTOBER 14, 2025

On Friday, October 10, the normally understated *The Wall Street Journal* described the U.S. budget picture in one word: grim. The United States is consistently running federal deficits over 6 percent of GDP—levels historically seen only during wartime or deep recessions.

After months of budget cutting, rising tariffs, and solid GDP growth, the 2025 federal deficit, rather than falling, increased. According to the Journal, the Department of Government Efficiency (DOGE) was "not even close" to cutting $2 trillion in federal spending as Elon Musk had initially promised. The paper blamed the nation's deteriorating federal budget on "rising social programs, including Social Security, and Medicare, and interest on the public debt, which topped $ 1 trillion by one measure for the first time."

The financial markets apparently agree with *The Wall Street Journal's* grim assessment. Since the beginning of the year, the dollar has declined 10 percent relative to other leading currencies. Gold and Bitcoin, hedges against the dollar, are near all-time highs. Foreign investment, and trust, in U.S. Treasury bonds are declining.

Yet, the Federal Reserve is still able to finance the nation's growing debt at reasonable interest rates; U.S. Treasury bonds are currently yielding only 4.2 percent. In spite of the nation's large federal deficits, the U.S. dollar remains a "safe haven" for the world's money managers. But that can quickly change as the below chart shows. After the U.S. government lost control of inflation during the nineteen-seventies, Treasury bond interest rates—in order to entice buyers—rose to over 15 percent in 1981. A similar scenario today would be catastrophic. Even a doubling of bond interest rates would increase the annual cost of financing the national debt to over $2 trillion.

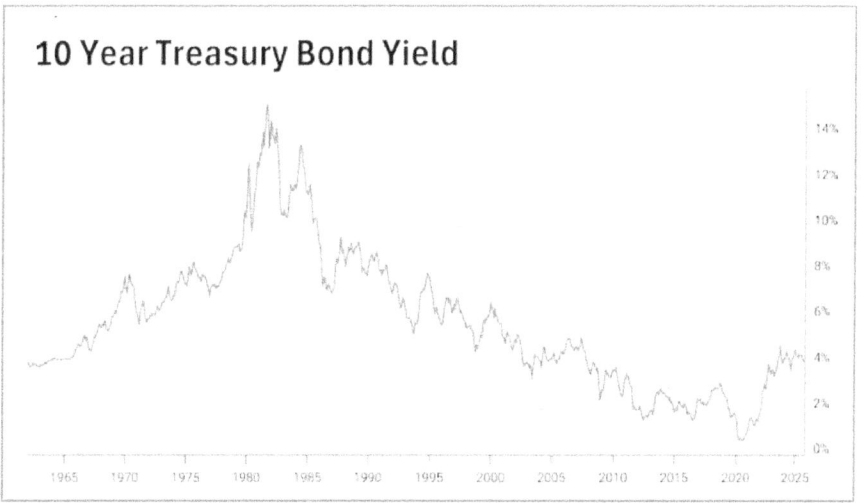

U.S. FEDERAL RESERVE

Of course, the worst scenario would be the collapse of the U.S. dollar. After eighty years as the world's reserve currency, that might seem inconceivable. But as a fiat currency, the only underlying value of the dollar is trust. Trust that the United States government will prudently manage its finances, trust that inflation will be kept under control, and trust that America will remain stable politically.

*The Wall Street Journal* article only touched on the factors threatening the U.S. budget. The rising deficit is more serious than even the Journal suggested.

## The United States has the world's highest per capita healthcare costs

The United States is the only major developed nation that does not have a national healthcare plan. Instead, U.S. healthcare is a hodgepodge consisting of private insurance plans, Medicaid, Medicare, and the Veterans Administration (VA). About 8 percent of Americans, 27 million, have no health insurance.

Compared to other developed countries, Americans have poor health outcomes, and the shortest lifespans.

Forty-two percent of Americans are obese compared to other developed countries such as Canada (28 percent), United Kingdom (27 percent), Germany (23 percent), Italy (18 percent), and France (10 percent). Americans die of coronary disease—73.5 deaths per 100,000 population—far more often than other countries: Canada (45.2), United Kingdom (43.1), Germany (56.1), Italy (42.4), and France (30.2). America's maternal death rate—17 per 100,000 live births—is shameful compared to other countries: Canada (12), United Kingdom (8.3), Germany (3.6), Italy (6.5), and France (7.3). During the Covid pandemic, the United States, by a large margin, had the highest death rates of any of the twenty largest developed nations (excluding China due to unreliable reporting).

So, even if the U.S. government managed its healthcare budget very well, federal healthcare spending would remain high since America's fractured healthcare system, compared to other developed nations, is more costly and also less effective in keeping Americans healthy.

Unfortunately, restructuring the nation's healthcare system is very difficult; Congress is showered by money from the healthcare lobby. In 2024 alone, healthcare-related companies spent $742 million—$1.4 million for every member of Congress—promoting their interests.

## The U.S. population is rapidly aging

In 1938, Social Security was designed as a "pay-as-you-go" system in which current workers' payroll taxes fund current retirees' benefits. That worked well for many decades. In 1950, sixteen workers supported each Social Security retiree. That's not surprising, in 1950 the average lifespan was only 68.1 years. But modern medicine has steadily increased lifespans and a child born today can expect to live nearly 80 years.

Yet, since 1938 Social Security's Full Retirement Age (FRA) has only increased by two years, from 65 to 67 years. Had Social Security regularly raised the FRA as lifespans lengthened, Americans today would only become eligible for Social Security benefits in their mid-seventies.

In 2000, well before Baby-Boomers began to retire, Social Security payroll taxes more than covered expenses: $1.20 for every dollar paid in benefits. Today, payroll taxes only cover about 85 percent of Social Security's costs. The balance is financed by selling Treasury bonds which contributes to our rising national debt.

Other than President Reagan's intervention in 1983 (see my October 7, 2025 Substack), Congress has done little to balance the Social Security budget. Worse, rather than just "kicking the can down the road," Congress has steadily increased Social Security benefits even as its deficits ballooned. This year, for example, the "One, Big, Beautiful Bill" Act approved a $6,000 tax deduction for eligible Social Security recipients.

## U.S. defense spending is likely to increase—perhaps significantly

Today, the United States far outspends the world in defense spending as the chart below shows. With a defense budget approaching $1 trillion, U.S. defense spending in 2024 exceeded the combined defense spending of the next nine countries combined.

But over the last few years, global tensions have risen to their highest level since the end of the Cold War. Russia seems to have no interest in ending the Ukraine War, China is rapidly expanding its military spending, and hope for peace in the Middle East remains fragile. Yet, U.S. defense spending, measured as a percentage of GDP, is near its lowest level since before the 9/11 Terrorist Attacks in 2001.

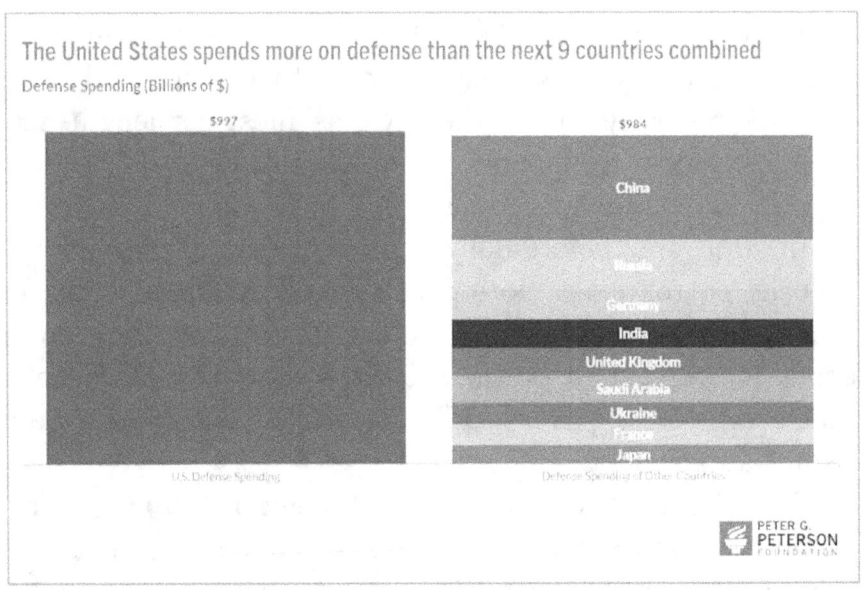

Today, China's defense spending, measured as a percentage of GDP, is half the United States'. That difference is rapidly shrinking. Long a second-rate military power, China is expanding its Pacific naval fleet, building advanced stealth fighters, upgrading its missile systems, and launching military and intelligence satellites.

So, it is very possible that the United States will be forced to significantly increase its defense spending to maintain par with China.

## Congress seems incapable of dealing with the federal deficit

A quarter century ago, the United States enjoyed a period of extraordinary fiscal strength and optimism. From 1998 through 2001 the United States ran strong budget surpluses. The Congressional Budget Office (CBO) was so confident that the surpluses would continue that it projected the federal debt would fall from 36.1 percent of GDP in 2001 to 6.3 percent in 2010.

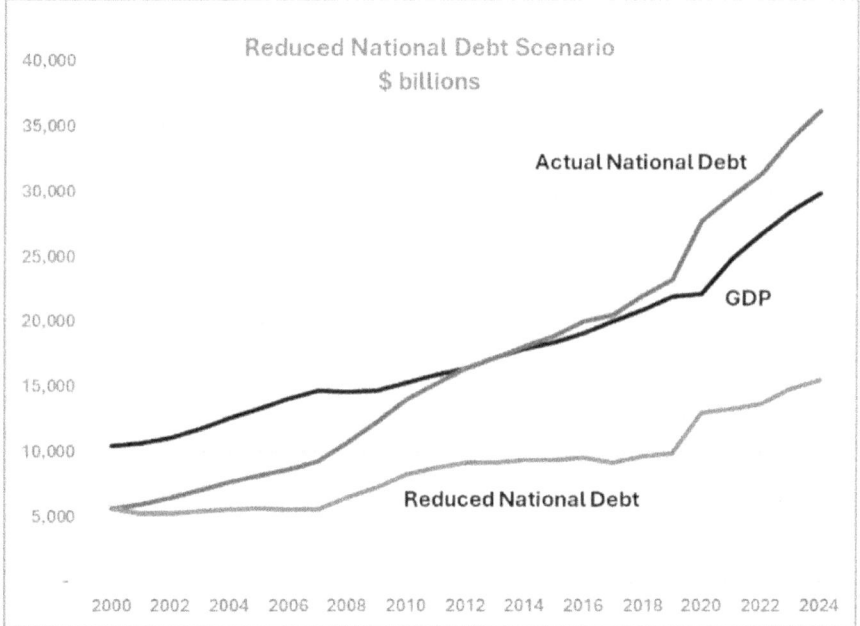

Had Congress in 2000 limited the growth of the national debt to GDP growth, today's national debt would be $16 billion rather than $36 billion.   FEDERAL RESERVE

That, of course, did not happen. The 9/11 Terrorist Attacks pushed the country into two Middle East wars, President Bush cut taxes while increasing healthcare spending, the 2008 Financial Crisis drove the nation into a deep and sustained financial slump, the Covid pandemic generated the largest federal deficits since the Second World War.

By 2025, the U.S. national debt had soared to 120 percent of GDP.

But what would the national debt be today if Congress had somehow limited the growth of the federal deficit to that of the GDP—which from 2000 through 2025 averaged a healthy annual growth rate of 4.5 percent? The above chart answers that question.

Had a responsible Congress starting in 2001 constrained federal deficits to grow no faster than the national economy, the national debt today would be approximately $15 trillion, about half of GDP, and far lower than today's $36 trillion national debt.

The needed reductions in healthcare and Social Security deficits would have been painful. Whether done by spending cuts, tax increases, or a combination of both, federal healthcare deficits (Medicare, Medicaid, and the Veterans Administration) would have been reduced by 40 percent, Social Security deficits by 20 percent, and all other deficits, including defense, by 10 percent.

In addition to reducing services and cutting wasteful spending, any significant reduction in the federal deficit will require an increase in taxes. Americans enjoy the lowest taxation rates of any major developed country other than Ireland. But those low taxation rates have been at the cost of a soaring national debt.

Today, slashing federal deficits seems unobtainable. But if the United States is ever confronted with a fiscal crisis similar to other countries that failed to manage their spending (Mexico in 1994, South Korea in 1997, Russia in 1998, Argentina in 2001, Iceland in 2008, Turkey in 2018), we Americans will be asking ourselves why didn't Congress, and the president, somehow find the resolve to bring the federal budget into line when it was still possible?

# A Brief History of Artificial Intelligence: Part 3

OCTOBER 21, 2025

A popular game show, Let's Make a Deal, has a common root with today's Artificial Intelligence (AI) systems.

The show has several variants in which audience members, known as "traders," negotiate for prizes. One popular variant is the Monty Hall Problem after the show's original host. In this game, the trader is presented three doors. Behind one door lies a valuable prize such as a new car or resort trip. Behind the other two doors are gag prizes, known as "Zonks," such as a goat.

The game starts when the host asks the trader to pick a door. The door remains closed. The host then opens one of the other two doors which he already knows has a Zonk behind it, leaving two closed doors. Behind one of the closed doors is the prize, behind the other is a Zonk. The host then asks the trader, "Do you wish to switch your selection to the other door?"

What would you do?

―――

Sometime around 1750, an English minister named Thomas Bayes became interested in probability theory, a new branch of mathematics at the time. Bayes was particularly interested regarding how to predict an event for which no information is initially available—exactly the case for the Monty Hall Problem.

To help him analyze his theory, Bayes relied on a fictional experiment. He is standing with his back facing a large table when he asks an assistant to toss a ball on the table. With his back turned, Bayes does not know where the ball landed. So he takes a guess. He then asks his assistant to randomly toss a second ball onto the table and then asks the assistant to tell him whether the ball landed to the left or the right of the original ball. Based on this new information, Bayes can refine his original guess since a ball landing to the left, say, of the original ball suggests the original ball is not near the left edge of the table.

The Monty Hall Paradox.                      SCIENTIFIC AMERICAN

Bayes then asks his assistant to toss the ball repeatedly, each time giving Bayes slightly more information regarding where the original ball is located on the table. Given enough ball tosses, Bayes can determine the location of the original ball.

That was Thomas Bayes' insight. Even when no information is initially available, complex problems can be solved by conducting a series of trials, each of which yields information pointing to the solution.

Bayes' discovery was published posthumously in 1761. Decades later, the French mathematician Pierre-Simon Laplace refined and formalized Bayes' work giving Bayes full credit. But for over a century, Bayes Theorem largely lay dormant; mathematicians resisted a theory which was based on a whimsical guess as a starting point.

The Second World War changed that. Every month, German submarines were sinking hundreds of Allied ships in an effort to starve Britain into surrender. The Allies were desperate to break the German Enigma codes which the German Naval Command used to communicate with its submarines at sea.

British mathematician Alan Turing realized the problem was perfect for Bayes Theorem. But to crack the code, untold thousands of successively better guesses would need to be made. After two years of intense work, Turing and his team developed an electro-mechanical computer to automate the process. By 1941, the British were routinely decoding Germany's top-secret messages in under an hour. (For an account of how Turing broke the German codes, see the movie *The Imitation Game*.)

Winston Churchill later told King George VI breaking the German codes was responsible for the Allies winning the war.

But what has all this to do with Let's Make a Deal? And what about your decision? After the host opened a door, would you stick with your initial selection or switch?

Most people would agree if you stuck with your first decision. There are two remaining closed doors so "common sense" suggests each door has a 50 percent chance of having the prize behind it. So why switch?

But you would be wrong. By switching doors, you would double your chances of winning the prize. Here's why:

1. At the beginning of the game, the door you pick has a one-third chance of having the prize behind it. The other two doors, together, have a two-thirds chance.
2. After the host opens one of the other two doors to reveal a Zonk, you know the prize is either behind your pick or the remaining unopened door.
3. Now here is the confusing part. Since your original pick had a one-third probability of winning the prize, that probability does not change after the host opens one of the two other doors. Instead, the remaining unopened door now has a two-thirds probability that the prize is behind it.

Skeptical? That's understandable, the solution is counter intuitive. So counter intuitive that in 1991 *The New York Times* wrote a major article on how the Monty Hall Problem confused some of the world's top mathematicians.

But play the game with a deck of cards and the solution is more obvious.

1. The host lays out the fifty-two cards face down and asks you to pick the Ace of Spades. You pick a card. It only has a 1 in 52 chance of being the Ace. The other fifty-one cards, together, have a 51/52 chance of having the Ace.
2. The host turns over fifty of the fifty-one cards you did not pick. None are the Ace of Spades.
3. The host then asks if you wish to switch your selection. Of course, you want to switch! The card you originally picked only had a 1/52 chance of being the Ace of Spades. That hasn't changed. But the other face-down card now has a 51/52 chance of being the Ace of Spades—the same probability as the fifty-one cards you did not originally pick.

A mathematician relying on Bayes Theorem would easily have solved the Monty Hall Problem. Unlike traditional statistics, Bayesian statistics consider how probabilities change as a situation changes. During Let's Make a Deal, only one event occurs, a door is opened. Yet that event significantly changes the game's probabilities.

Similar to Let's Make a Deal, Artificial Intelligence uses Bayes Theorem to evaluate probabilities after changing a situation. But rather than a single change, AI makes millions, even billions of changes while successively evaluating the effect of each change.

Suppose you ask your iPhone, "Siri, play three of Bob Dylan's most popular songs." The AI system first segments this sentence into "tokens," discrete fragments that might represent whole words, word fragments, individual characters, or punctuation. The tokens are then converted to numbers for computer processing.

Once tokenized, the AI system must identify "Bob Dylan" as a specific artist. The system, using Bayes Theorem, has already learned from vast internet datasets that when the tokens "Bob" and "Dylan" appear together in music contexts, they almost certainly reference the musician. During its training, the AI system has also learned, and memorized, tokens that are suggestive of music. It searches for these tokens in the sentence. Finding the token, "songs," the system concludes that there is a high probability that the sentence is related to music. A similar process is followed to interpret the meaning of the other tokens within the sentence.

It is by this simple, incremental process—like tossing a ball on Bayes' table—that Artificial Intelligence systems learn, understand, and reason.

For millennia, the ability to reason was thought to be exclusive to humans. "I think, therefore I am," philosopher René Descartes famously declared. But in 1949, Alan Turing questioned that assumption when he asked in a landmark paper, "Can machines think?"

Turing's paper laid out a thought experiment known today as the Turing Test but called "The Imitation Game" by its inventor. The Game was simple: a human participant would exchange a series of typed interactions with two respondents, a computer and a human being. Each respondent—one of flesh, the other circuit-bound—remained hidden behind a partition. After a set period of time, if the interrogator failed to distinguish one from the other, the computer would, in effect, win; such a machine could be said to think.

Can machines think? Today, Alan Turing would almost surely answer, "yes."

# Can Artificial Intelligence Repair America's Media Divide?

OCTOBER 28, 2025

An October 21 editorial in *The Wall Street Journal*, "Biased Media Needs a Complete Culture Change," inspired today's Substack.

"It isn't true that trust in media has simply collapsed across the board," the *Journal* editorial declared. "Instead, it has become partisan. Consumers trust only what they already believe."

This is old news. The Fairness Doctrine was repealed in 1987 during the Reagan administration.

After the Second World War, concerned that American broadcasters could become propaganda outlets like those in Nazi Germany and the Soviet Union, the FCC with Congressional support passed the Fairness Doctrine. The Doctrine required "broadcasters to present fair and balanced coverage of controversial issues of interest to their communities."

For nearly forty years, radio and television news—considered an essential public service— was widely regarded as being factual and trustworthy. Walter Cronkite, Chet Huntley, David Brinkley, Edward R. Murrow, and John Chancellor were some of the most respected men in America. The FCC called the Fairness Doctrine the "single most important requirement of operation in the public interest – the *sine qua non* for grant of a renewal of license."

But after the Doctrine's repeal, news media were free to tailor the news to their target audience mixing facts, opinions, and misinformation. The strategy worked making news—once a money-losing public service—into a multi-billion-dollar industry.

During the 1960s and 1970s, CBS newscaster Walter Cronkite was considered "the most trusted man in America."   CBS NEWS

But trust in media collapsed. According to Gallup polls, in 1975 over 70 percent of Americans had a great deal or fair amount of trust in the nation's news media. Today, less than 30 percent feel the same way.

Together, partisan politics and the profit-based news media have turned America into deeply divided Red and Blue States. This is not healthy for our Republic.

But it is just possible that Artificial Intelligence can restore the Fairness Doctrine—for those who want it. Today's AI can instantly read hundreds of news sources to, in theory, provide a balanced, factual perspective.

As a simple experiment, I asked three leading AI systems—chatGPT, Claude, and Grok—to report on the October 18 "No Kings" protests. I made no changes to the results.

**My Request**

Write a 300-word article, including a headline, in the style of a newspaper reporting on the "No Kings" protests held across the nation on October 18, 2025.

**chatGPT Responded**

### Nationwide "No Kings" Protests Demand New Vigilance for Democracy

In cities large and small across the United States, Saturday's nationwide mobilization under the banner No Kings protests carried an unmistakable message: the people are rising to challenge what they view as creeping authoritarianism. The day of action, held Oct. 18, 2025, brought together millions of demonstrators in more than 2,600 locations, from the heart of downtown Washington, D.C. to regional squares in the Midwest.

Organizers say the movement—branded to emphasize that America "has no kings"—is a response to what they describe as an accumulation of executive power under Donald Trump, and erosion of democratic norms, including concerns over civil-liberties crackdowns, militarized policing, and election integrity. At major rallies in New York, Chicago and Washington, protesters bore banners quoting the Constitution, wore bold costumes and inflatable figurines, and marched in vibrant, spirited fashion.

Despite fears of clashes, law-enforcement presence remained largely local, and the day passed with remarkably few arrests in major cities. In San Diego County, for instance, more than 25,000 took part with no arrests, according to the police department. Republican leaders were largely critical or silent, branding the protests as "hate-America rallies," but many demonstrators claimed the opposite: they were acting out of love for the country and its democratic heritage.

As the sun set on this intensely visible nationwide event, the overarching question remains: will "No Kings" be remembered as a show of solidarity, or will it translate into sustained political change? The movement's leaders say this is only the beginning—and that today's protest is a warning, not a finale.

## Claude Responded

### 'America Has No Kings': Massive Protests Sweep Nation in Challenge to Trump Policies

Demonstrators filled city streets from coast to coast on Saturday as an estimated seven million Americans participated in "No Kings" protests, the second major coordinated action against President Donald Trump's administration this year.

The protests, organized by a coalition of progressive groups including the ACLU, Indivisible, and MoveOn, took place at more than 2,700 locations nationwide. Participants rallied against what they described as authoritarian governance, with particular focus on immigration enforcement operations, military deployments in Democratic-led cities, and the ongoing government shutdown.

"Together, millions will send a clear and unmistakable message: we are a nation of equals, and our country will not be ruled by fear or force," organizers said in a statement.

The day's events were largely peaceful. In New York, thousands packed Times Square, while Chicago protesters waved upside-down American flags alongside "Hands Off Chicago" signs. Los Angeles saw at least ten separate gatherings featuring salsa bands and demonstrators in inflatable costumes. Senator Bernie Sanders addressed crowds in Washington, and an estimated 15,000 people rallied at Utah's State Capitol.

Political tensions flared over the demonstrations. House Speaker Mike Johnson called the protests "a hate America rally" and claimed they were preventing government reopening negotiations. Democratic leaders pushed back sharply, with organizing groups calling Johnson's remarks an attempt to "demonize peaceful protest."

Several Republican governors deployed state resources ahead of the rallies. Texas Governor Greg Abbott mobilized National Guard units, state troopers, and tactical assets to Austin, while Virginia's Governor Glenn Youngkin activated Guard members for training purposes.

The October mobilization follows June's inaugural "No Kings" protests, which drew approximately five million participants nationwide in opposition to a military parade coinciding with Trump's birthday.

**Grok Responded**

### 'No Kings' Protests Unite Millions in Historic Push Against Trump Policies

Across the United States, on October 18, 2025, an unprecedented wave of protests dubbed "No Kings" swept through over 2,700 locations, from urban centers to rural towns, in a vibrant stand against President Donald Trump's policies. Organized by a coalition of 200 progressive organizations, including the Women's March and Indivisible, the demonstrations drew an estimated 7 million participants, making them the largest single-day protest in U.S. history.

In Washington, D.C., 300,000 rallied at the National Mall, waving signs reading "Democracy, Not Dictatorship." New York City saw 100,000 marchers in yellow, symbolizing resistance, while Los Angeles' 200,000-strong crowd filled Pershing Square. Smaller cities like Asheville, North Carolina, with 15,000, and Fargo, North Dakota, with 8,000, joined the chorus. Protesters, some in playful costumes like inflatable dinosaurs, sang Sweet Caroline with rewritten lyrics mocking Trump's "kingly" ambitions.

The protests targeted Trump's recent executive actions, including expanded immigration crackdowns, environmental regulation rollbacks, and federal interventions in blue states. In Atlanta, 150,000 heard Rep. Stacey Abrams warn, "Our votes, our voices, will not be silenced." Chicago's 250,000-strong rally featured Gov. JB Pritzker condemning "billionaire-backed tyranny." No major violence was reported, though tensions flared in Dallas, where counter-protesters clashed verbally with the 50,000-strong crowd.

Trump dismissed the protests on X, calling them "loser riots," while GOP allies like Sen. Ted Cruz labeled them "radical stunts." Yet, the peaceful, diverse turnout—from Anchorage's 5,000 to Miami's 80,000—underscored a unified message. "We're here to protect our future," said Maria Lopez, a Tucson teacher marching with her students. As the nation grapples with deepening polarization, the "No Kings" movement signals a formidable opposition, ready to challenge what they see as democratic erosion.

# Don't Cry for Me Argentina

NOVEMBER 4, 2025

Last month, President Trump authorized a $20 billion financial lifeline for Argentina.

Why Argentina? A country most Americans know largely through the musical and 1996 movie, *Evita*, its passionate tango dancers, and Argentine wine and beef.

But Argentina has a rich and, unfortunately, often tragic history.

Like the United States, starting around 1870 Argentina experienced its own gilded age. British money flowed into Argentina to finance infrastructure, particularly railroads spanning the Argentine Pampas whose vast open plains and rich soil made it one of the world's top agricultural regions. By 1890, Argentina's annual GDP per capita growth exceeded that of the United States and Great Britain. The immense wealth transformed Buenos Aires from a pretty little colonial town to a beautiful city with tree-lined boulevards, French and Italian architecture, and hundreds of parks. By 1913, Argentina's economy was on a par with France and Germany. For decades, the French used the expression, *riche comme un Argentin* — rich as an Argentine.

It didn't last. Argentina's export-based economy collapsed during the early days of the Great Depression. Import tariffs erected by the United States and Europe slashed Argentina's beef and grain exports, the country's lifeblood. Argentina quickly fell to military dictatorships which turned the nineteen-thirties into Argentina's "Infamous Decade," years of electoral fraud, persecution of political opposition, and widespread government corruption.

In 1946, the military dictatorships ended when Juan Peron was elected president. Peron was a shrewd populist whose political base was composed of trade unions and their disenfranchised workers.

Elon Musk and Argentine President Javiel Milei at the Conservative Political Action Conference in February 2025.     FACEBOOK

Peron established a central bank, nationalized foreign companies, implemented broad wealth distribution, expanded social welfare programs, and created a system of free universal education. Assisted by his popular and engaging wife, Eva, Peron remained in power until 1955. That year Peron was overthrown by yet another military dictator.

For the next seven decades, with few exceptions, Argentina suffered under either cruel military dictatorships responsible for "disappearing" as many as 30,000 "enemies of the state," or populist governments whose misguided economic policies wrecked the economy.

Argentine President Javier Milei just might be different.

Here in the United States, Javier Milei is probably most famous for the photo of him with Elon Musk brandishing a chainsaw at the Conservative Political Action Conference (CPAC) in February 2025. For many Americans, it's the image of two daffy jokesters. But at least for Milei, that is hardly the case.

Similar to Donald Trump, Javier Milei was a media star in Argentina before his rise in politics. He was also an accomplished economist long critical of the government's economic policies. Milei became famous for his humorous and acerbic mocking of Argentina's political establishment which he calls La Casta, or The Political Caste—much as Donald Trump refers to federal bureaucrats as The Deep State.

If Milei had a campaign slogan, it was *¡Que se vayan todos!* — They all must go!

After serving just two years in Argentina's Chamber of Deputies (similar to the U.S. House of Representatives), in October 2023 Milei defied the polls and won the presidential election with a record 56 percent of the popular vote.

Milei made three campaign promises: (1) to "take a chainsaw" to Argentina's political establishment slashing government bureaucracy, spending, and regulation, (2) balance the budget, and (3) end runaway inflation.

During his first year as president, Milei astonished his detractors by largely delivering on his first two promises. As the January 18, 2025, *Buenos Aires Times* reported:

> "Milei has applied shock therapy to try to stabilize South America's long-struggling second-biggest economy.... laid off more than 33,000 public sector workers, slashed state subsidies on transport, fuel and energy and led a massive deregulation drive.... [and] halved annual inflation, which fell 94 points to 117.8 percent last year.... "The promises

have been kept. The 'zero deficit' is a reality... Milei wrote on Instagram."

To balance the 2024 budget, Milei cut expenditures, excluding interest payments, by 32 percent. It was a remarkable accomplishment, both for Milei to deliver the bold plan and Argentine society to accept the deep spending cuts which briefly pushed poverty levels to over 50 percent.

But a 117.8 percent annual inflation rate remains a serious problem. Milei's answer to Argentina's chronic inflation is to replace the Argentine peso with the U.S. dollar. For this, Milei is likely looking to another South American country as a model.

In 1999, after years of reckless spending, the Ecuadorean sucre lost two-thirds of its value pushing inflation up to 96 percent. The next year, Ecuador abandoned the sucre and adopted the dollar as its national currency. No longer able to print money, Ecuador was forced to impose fiscal discipline and reduce government deficits. It worked. Today, Ecuador's inflation rate is under 1.0 percent.

Similarly, Milei has little choice but to abandon the Argentine peso. For decades, Argentina defaulted so often on its financial obligations that the peso is shunned worldwide. Unable to borrow, Argentina was forced to "print money." Argentina's central bank simply wrote checks for money it did not have. This increased the supply of money circulating through the economy leading to even more inflation—when more money chases the same amount of goods and services, prices inevitably rise.

Unfortunately, inflation has worsened since Milei became president. Since December 2023, the peso has declined from 700 pesos to the dollar to nearly 1500 pesos today. Ultimately, this level of runaway inflation destroys a country's economy.

On October 15, President Trump stepped in when the U.S. Treasury announced it would purchase $20 billion worth of

Argentine pesos. In addition, Treasury Secretary Scott Bessent announced the administration was arranging an additional $20 billion rescue fund from the private sector. These large purchases of pesos not only reduce the supply of pesos in the Argentine economy, but also telegraph American support for Milei's policies.

The announcement stunned Trump's supporters. How does bailing out Argentina square with America First? Just months before, the president proudly announced that the United States had slashed humanitarian aid—food, shelter, medicine, and other life essentials—to some of the world's poorest countries. (In 2024, U.S. humanitarian aid totaled $15.3 billion—twelve cents a day on a U.S. per capita basis.)

As CNN reported, President Trump justified the investment because "having a likeminded and enthusiastic [U.S.] ally in Buenos Aires has been viewed as a boon. The country has large deposits of key minerals, like lithium and copper, that are critical to US manufacturing." But ultimately, the bailout was due to President Trump's strong support for Milei. "He's MAGA all the way," Trump said. "It's Make Argentina Great Again."

We don't know how President Trump's intervention in Argentina will turn out. Let's hope it will be as successful as President Clinton's intervention in Mexico during January 1995. When Mexico faced unexpected financial instability, the United States committed to a $20 billion bailout package through the International Monetary Fund.

Mexico only used $13 billion of the package and quickly paid back the loan generating a $600 million profit for the U.S. Treasury. For the next thirty years, Mexico has remained a stable and valued trade partner.

Let's wish the same for Argentina.

# America's Neglected Air Traffic Control System

NOVEMBER 11, 2025

Is there a profession more critical to public safety than our nation's air traffic controllers (ATC)? I doubt it.

Police and firemen protect our lives but many, perhaps most, Americans live their lives fortunately never needing their protective services. Our Armed Services protect the nation, but over the last century only two attacks have been made by foreign invaders on American soil: the Japanese attack on Pearl Harbor on December 7, 1941, and the September 11, 2001, terrorist attacks.

But every day, U.S. air traffic controllers direct the flights of more than 44,000 aircraft flown by pilots ranging in experience from veteran airline pilots to student pilots on their first solo flight.

Yet, mid-air collisions between aircraft are vanishingly rare. In the last twenty-five years—after an estimated 1.2 million flights—only three accidents occurred in the United States in which controller error was the primary cause. All involved small, private aircraft. (The tragic January 29, 2025, mid-air collision between American Airlines Flight 5342 and a U.S. Army Black Hawk helicopter over the Potomac River in Washington, D.C. remains under investigation; early evidence points to a combination of pilot error and ATC understaffing.)

It all started in 1929, with Archie League. That year, the St. Louis Lambert Airport hired Archie to direct aircraft. After Charles Lindbergh's historic 1927 transatlantic flight, interest in aviation was booming making the pilots' simple rule of "see, and be seen" increasingly dangerous. Aviation needed their own traffic cops.

Archie League in 1929. America's first air traffic controller.　　FAA

Archie's "control tower" was a wooden wheelbarrow which held a lawn chair, beach umbrella, his lunch box, and two large flags: a checkered flag and a red flag. Based on the wind, Archie would roll his wheelbarrow to the end of the active runway. There he signaled departing and landing aircraft using the two flags. The checkered flag signaled to proceed, the red flag to hold and await further instructions.

But technology was developing rapidly. In 1930, the Cleveland, Ohio airport established the first radio-equipped airport tower greatly expanding the communications between the ground and aircraft.

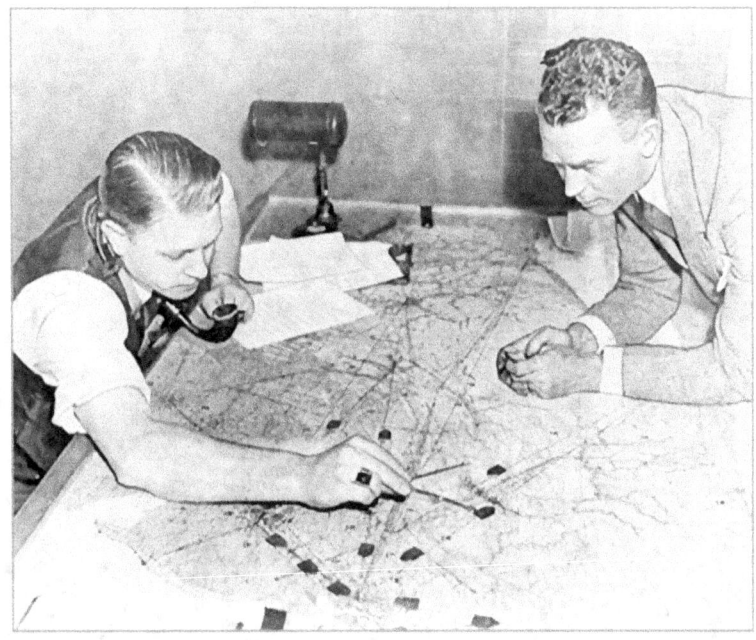

Air traffic controllers using "shrimp boat" markers to track aircraft circa 1945.
FAA

In December 1935, the first Airway Traffic Control Station was established at Newark, New Jersey. As pilots radioed in their position, controllers moved plastic markers, known as "shrimp boats," across a large horizontal map showing the location of aircraft. This simple system allowed controllers to not only control aircraft around airports, but also as they flew enroute reducing the risk of midair collisions.

Aviation expanded tremendously during the Second World War. As the war ended, twenty-seven air traffic control "centers' were controlling aircraft across the nation. Although mid-air accidents happened, they typically involved small private aircraft.

By the standards of the nineteen-thirties, commercial aviation had become relatively safe.

Until June 30, 1956. That morning, two flights departed from Los Angeles International Airport only minutes apart: TWA Flight 2, a Lockheed Super Constellation, bound for Kansas City with seventy crew members and passengers and United Flight 718, a Douglas DC-7, heading to Chicago with fifty-eight people aboard.

The horrendous 1956 airline crash over the Grand Canyon forced the United States to modernize its ATC system.

Ninety minutes later, the two aircraft collided over the Grand Canyon. The DC-7's left wing clipped the top of the TWA Super Constellation's vertical stabilizer causing the tail assembly to break away from the rest of the aircraft. The TWA Constellation plunged in a vertical dive into the canyon. The

United DC-7 managed to stagger a mile before it slammed into a mountain.

All 128 people on board both aircraft died, making the Grand Canyon crash the first commercial airline accident to exceed one hundred fatalities. The accident shocked the world.

How could two modern airliners, flown by experienced crews, somehow collide in clear weather? Actually, it is easy to miss another aircraft on a converging course. The converging aircraft, although slowly increasing in size, appears stationary against the background until it's suddenly too close to avoid. Sun glare, task saturation, and complacency exacerbate the risk. That's why private pilots, even in clear weather, typically ask ATC to provide "flight following" to advise them of nearby aircraft.

The Grand Canyon accident fundamentally changed air traffic control in the United States. In 1958, the Federal Aviation Agency (FAA), with far more regulatory authority, replaced the Civil Aeronautics Administration. Although it would take decades, the FAA transformed air traffic control from a loosely regulated system based on moving markers on horizontal maps to today's radar-based ATC system with pilots in constant contact with air traffic controllers.

Today, even before the current Congressional shutdown, the FAA air traffic control system has become a political football. Torn by competing interests, much of the nation's ATC system is obsolete, understaffed, and underfunded. Political considerations complicate ATC management. Congress often directs FAA spending based on constituent interests rather than true operational priorities such as upgrading computer and radar systems. Rather than being driven by an open market, controllers' pay scales are subject to congressional oversight.

Chronic understaffing forces controllers to often work a compressed work schedule ominously known as the "rattler" where air traffic controllers work five eight-hour shifts crammed into four days. The schedule includes two consecutive afternoon/evening shifts ending at 10:00 pm, followed by two consecutive morning shifts starting at 7:00 am, and concluding with a single overnight shift starting at 10:00 pm.

Cynical politics even affects safety. In one infamous incident, Senator Jim Inhofe from Oklahoma in 2010 landed his Cessna 340 aircraft on a closed runway marked by a large "X," the international symbol for a closed runway. During the FAA investigation, one terrified worker "describe[d] how Inhofe's plane touched down on the runway then 'sky hopped' over the six vehicles and personnel on the runway, and then landed." Normally, such a violation would result in a license suspension at a minimum. But Inhofe was let off easily by the FAA with little more than a wrist slap. Incredibly, Inhofe later sponsored a bill to weaken FAA enforcement of safety regulations.

Unlike the United States, other countries including Australia, Canada, Germany, and the United Kingdom have taken steps to remove their ATC systems from political manipulation.

Airservices Australia is a government-owned corporation responsible for providing air traffic control and other aviation services. Although not a private company, it operates under a commercial framework.

Similarly, Germany's ATC system is the government-owned company *Deutsche Flugsicherung* (German Air Safety) which operates under private law as a commercial enterprise.

The United Kingdom's National Air Traffic Services is a public-private partnership. The government retains a 49 percent stake with the remaining shares held by a consortium of airlines.

Canada, though, fully privatized its air navigation services in 1996 creating NAV CANADA. It is fully funded by fees charged to aircraft operators and receives no government funding. Many consider NAV CANADA the global standard for a well-managed ATC system.

For decades, starting with President Clinton in 1994, there have been efforts to privatize the U.S. ATC system. As recently as 2017, President Trump announced his intention to privatize the nation's ATC system. But like presidents before him, Trump was quickly dissuaded by powerful aircraft owner lobbies such as the Aircraft Owners and Pilots Association (AOPA), the Experimental Aircraft Association (EAA), and the National Business Aviation Association (NBAA). Private aircraft owners are concerned that a commercial ATC system would increase their costs and also reduce services at small, private airports.

Yet, America's 11,500 air traffic controllers soldier on working long hours in understaffed control centers with outmoded equipment. And today, during the Congressional shutdown, are even working without pay—while Congress shamefully continues to pay themselves.

# Immigration: Culture versus Economics

NOVEMBER 18, 2025

President Trump and Vice-President Vance have different views on immigration. Trump supports selective immigration, Vance believes America needs to get immigration "way, way down."

On November 11, during a Fox News interview with Laura Ingraham, the President defended H-1B work visas declaring the United States "has to bring in talent."

> Ingraham: "If you want to raise wages for American workers, you can't flood the country with tens of thousands or hundreds of thousands of foreign workers."
> Trump: "I agree, but you also do have to bring in talent."
> Ingraham: "Well, we have plenty of talented people in America."
> Trump: "No you don't. No."

President Trump cited the September raid on a Hyundai-LG battery plant located in Georgia as an example. "They had like 500 or 600 people, early stages to make batteries and to teach people how to do it," Trump said. "[Immigration agents] wanted them to get out of the country. You're going to need [them]."

Vice-President Vance disagrees believing the nation needs to slash immigration across the board. During an October Turning Point USA event at the University of Mississippi, Vance said, the optimal number of legal immigrants is "far less than what we've been accepting."

Vance believes immigrants are threatening the social fabric of the United States.

> "When something like that happens, you've got to allow your own society to cohere a little bit, to build a sense of common identity, for all the newcomers — the ones who are going to stay — to assimilate into American culture. Until you do that, you've got to be careful about any additional immigration, in my view... We have got to get our overall numbers way, way down."

President Coolidge and Vice-President Vance share similar views on immigration.　　　　　　　　　　　　　　　　　　　　WHITE HOUSE

Vice-President Vance's views on immigration echo those of President Calvin Coolidge a century ago. The Immigration Act of 1924, signed by President Coolidge, was an attempt to preserve a "common identity" by limiting immigration to Western and Northern Europeans. The Act replaced the Emergency Quota Act of 1921 passed under President Harding to reduce the influx of Eastern and Southern Europeans—namely, Jews, Poles, and Italians—after the First World War and the Russian Revolution.

President Coolidge, and most Americans, strongly supported the 1924 Immigration Act. During his 1924 presidential nomination speech, Coolidge declared:

> "Restricted immigration is not an offensive but purely a defensive action. It is not adopted in criticism of others in the slightest degree, but solely for the purpose of protecting ourselves. We cast no aspersions on any race or creed, but we must remember that every object of our institutions of society and government will fail unless America be kept American."

Coolidge's use of the phrase, America be kept American, isn't surprising. Four years earlier, Warren G. Harding had campaigned on "America First" after 117,000 American soldiers had died fighting during the First World War—a war many Americans believed the United States never should have entered.

Like Vance, President Trump believes in maintaining America's common identity. In 2018, during a meeting about immigration policy, Trump reportedly asked, "Why are we having all these people from shithole countries come here?" referring to immigrants largely from African nations. Instead, Trump suggested the U.S. bring in more people from countries like Norway, and this year expedited immigration of white immigrants from South Africa.

But Trump also understands U.S. immigration policy must consider economic factors. Multiple studies over the last two decades have concluded that immigration not only boosts economic growth, but also raises income levels for the majority of workers. In an April 2024 study, the National Bureau of Economic Research concluded:

> "Our estimates establish that immigrants have a substantial degree of productive complementarity with natives. This offsets the competition effect, resulting in a

boost of native wages and an increase of natives' employment-population ratio in response to inflows for most native workers."

Stated more clearly, qualified immigrants raise national productivity which benefits the majority of workers. This is particularly true in technology. Chinese and Indian workers, for example, make up only 6 percent of the total U.S. workforce but author or co-author 30 percent of all U.S. patents in strategic industries such as artificial intelligence, biotechnology, and communications according to studies. Their creativity boosts American productivity, and wages.

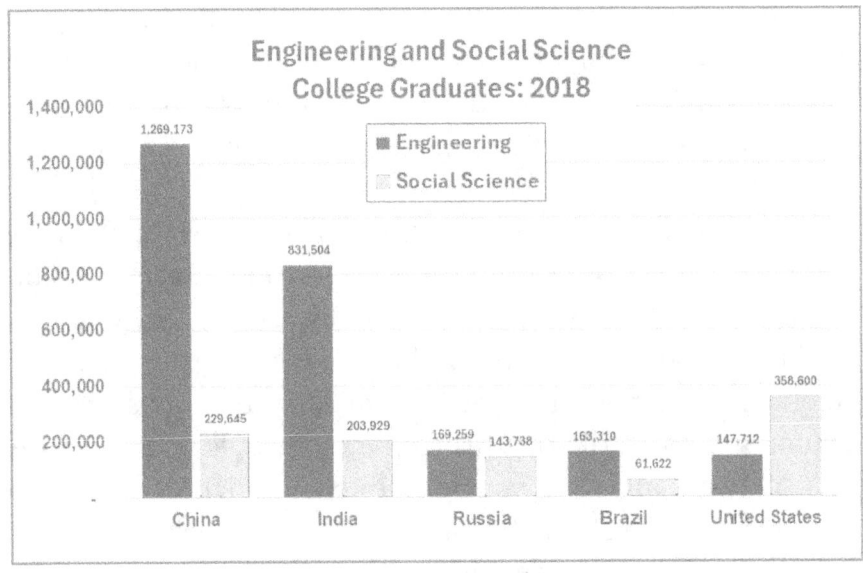

The United States greatly lags other countries in engineering graduates.
AMERICAN SOCIETY OF MECHANICAL ENGINEERS

Foreign-born scientists and engineers have long been critical to American industry. More than a century ago, two European immigrants, Nikola Tesla and Charles Steinmetz, developed the technology behind alternating current (AC) electrical systems used today around the world. Alexander Graham Bell, a Scottish immigrant, invented the telephone.

Guglielmo Marconi, while living in America, refined his invention of the radio and established the precursor to the Radio Corporation of America (RCA).

Even one of America's most ambitious scientific projects, the Manhattan Project to develop the atomic bomb during the Second World War, relied heavily on foreign-born scientists. Albert Einstein, Edward Teller, Felix Bloch, Victor Weisskopf, John von Neumann, Leo Szilard, Eugene Wigner, Enrico Fermi, Stanislaw Ulam, George Kistiakowsky, and other immigrants were key contributors to the massive project. Ironically, many of these scientists were Jews, Slavs, and Southern Europeans discriminated against two decades earlier by the 1924 Immigration Act.

More recently, America's early lead in semiconductors and software also owed much to immigrants such as Andrew Grove (Intel), Sergey Brin (Google), Jensen Huang (Nvidia), Charles Simonyi (Microsoft), and Elon Musk (Tesla and SpaceX).

But America's technology lead is eroding. Together, China and India produce 6.1 million STEM (Science, Technology, Engineering, and Mathematics) graduates annually compared to 820,000 in the United States.

The difference is even larger for engineers alone. China and India in 2020 graduated 1.7 million and 820 thousand engineers, respectively. The United States graduated a mere 126,000 engineers. Even Brazil and Russia graduate more engineers than the United States according to a 2018 study. In contrast, the U.S. produces more social science graduates than any other nation.

Either the United States needs to produce far more STEM graduates itself, or accept qualified, legal immigrants to supplement our nation's technical and scientific workforce.

One area where President Trump and Vice-President Vance both agree is the need to end birthright citizenship.

Since 1866, the 14th Amendment mandates that a child born on American soil, with few exceptions, is automatically granted U.S. citizenship. For more than a century, that interpretation was seldom questioned. America was a growing country and needed hard-working immigrants to fuel its economy.

Remarkably, until the 1924 Immigration Act, an immigration visa was not required to enter the United States. Except for certain groups such as Chinese, immigrants could simply board a ship and come to America. Upon arriving, agents determined who was admitted based on health and their judgement of the immigrant's character. (My father arrived in 1924 from Germany with only a letter of recommendation by his former employer.)

Once in the country, immigrants could become naturalized citizens, and their American-born children automatically granted U.S. citizenship. But today, should children born of parents residing in the United States illegally be granted U.S. citizenship?

The Trump administration says they should not. Most countries agree. The United States and Canada are the only two major developed countries that allow unlimited birthright citizenship. Other countries almost universally require that, at least, one parent be a citizen of the country.

In September, President Trump asked the Supreme Court to weigh in on the interpretation of the 14th Amendment, namely, "All persons born or naturalized in the United States, and subject to the jurisdiction thereof, are citizens of the United States."

Are illegal immigrants subject to the jurisdiction of the United States? That's the legal question the Supreme Court must answer.

# A Brief History of Artificial Intelligence: Part 4

NOVEMBER 25, 2025

If you were to ask people what company has most changed their daily lives over the last ten years, many would answer Amazon. By providing broad selections, low prices, and convenient home delivery, Amazon has changed how millions of people shop. (For others in local retail, it's been devastating.)

But what about the next ten years? What company will be most responsible for changing our lives and society? It might well be Nvidia. Nvidia has a near monopoly on the hardware and software—the infrastructure—that powers today's Artificial Intelligence systems such as ChatGPT, Claude, and Perplexity.

Nobody knows how AI will develop over the next decade. Pessimists believe AI's Large Language Model (LLM) technology is a dead-end since LLM is based on probabilistic models, what we humans call "guessing." Optimists believe AI will be deeply embedded in everyday life by 2030. Some believe AI will become self-aware, and ambitious.

Regardless, AI is broadly considered the most promising new technology since the development of computers seventy-five years ago, or even electricity a century ago. And today, Nvidia is the driving force in AI just as IBM was in computers and General Electric in electricity.

In addition to its strong technology lead, Nvidia is also a well-managed company. Nvidia controls over 80 percent of the global AI market. Unlike most growing technology companies, Nvidia is very profitable. Nvidia currently pockets $557,000 in net income for every million dollars in revenue. A $10,000 investment in Nvidia stock in November 2020 would be worth $140,000 today (November 24, 2025).

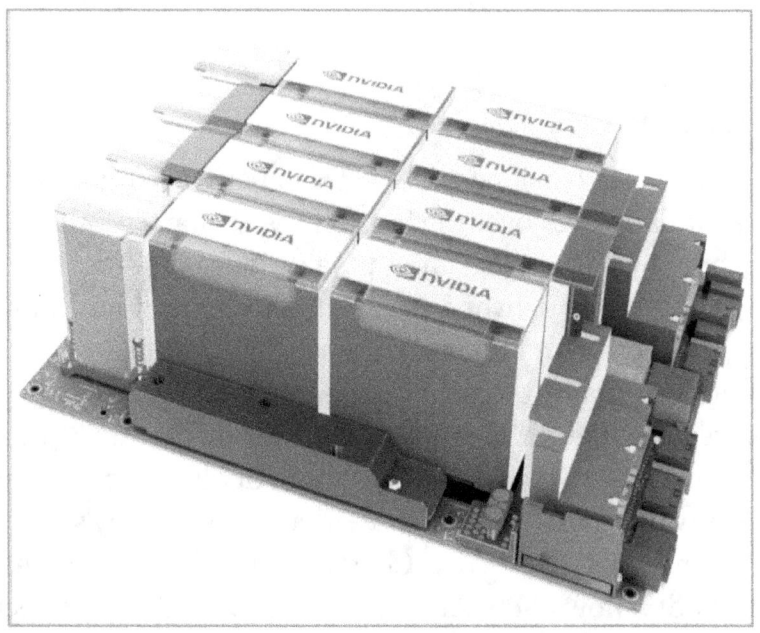

A fully configured Nvidia H100 system contains eight GPUs and can cost over $320,000.                                              NVIDIA

But Nvidia had a lucky break.

Nvidia was founded in 1993 to develop graphic accelerators to speed-up video games. At the time, video games were transitioning from simple two-dimensional images to realistic three-dimensional worlds. The new video games, such as Wolfenstein 3D and Doom, required much more computing power than available in personal computers and video game consoles at the time. Nvidia's graphic accelerators executed simple graphical calculations freeing the host processor to perform complex operations.

But in 1999, Nvidia released a new type of computing element known today as a Graphic Processing Unit (GPU).

Nvidia's GPUs transformed the video game industry, and would soon do the same for Artificial Intelligence.

Nvidia GPUs rapidly increased in performance making video games increasingly realistic. By 2010, Nvidia GPUs were approaching supercomputer speeds. That's when AI scientists realized the GPUs buried in their game consoles could be repurposed for AI applications. Nvidia had anticipated this and in 2007, ignoring the old management dictum to "stick to your knitting," had quietly announced an enhanced GPU designed for general-purpose processing, its "Compute Unified Device Architecture" (CUDA).

The breakout came in 2012. That year Alex Krizhevsky, a University of Toronto graduate student, along with Ilya Sutskever and their faculty advisor, Geoffrey Hinton, decisively won the ImageNet 2012 Challenge, an annual competition to encourage AI innovation. Their program, called AlexNet, was based on neural nets (see my July 22, 2025, Substack, "A Brief History of Artificial Intelligence: Part 2").

Alex worked from his bedroom at his parents' house using a computer supercharged with two Nvidia GPUs. AlexNet crushed the competition, beating finely-tuned software written by vision experts. It was just the beginning. Nvidia processors—now optimized for processing neural networks—would revolutionize research in Artificial Intelligence.

In 2016, Nvidia CEO Jensen Huang personally delivered the company's first AI supercomputer to the research lab OpenAI—an event considered by many to have kickstarted the modern AI revolution. Challenges that had stymied AI researchers for decades, from language translation to computer vision, were quickly advanced culminating in November 2022 with the release of ChatGPT—the first AI "killer app" which quickly garnered more than 100 million users.

Just why are GPUs, derived from video game processors, so well suited for Artificial Intelligence? Because both video games and AI are based on mathematical matrices that require massive amounts of processing.

The three-dimensional world inside a video game that looks so realistic and interactive for gamers is actually a mathematical matrix composed of billions of numbers. A typical representation might be a three-dimensional square matrix with 1,000 elements per side with ten numbers to represent each element. That works out to ten billion numbers just to represent one screen of a video game. A modern video game typically refreshes its screen 100 times a second. So if each element requires ten calculations to update it, the total calculations per second is something like one trillion ($1000^3$ x 10 x 100).

How big is a trillion? Nearly incomprehensible. Counting to a trillion at one second per count would require 31,700 years. Yet, a GPU does that and more every second.

Like video games, the "intelligence" in Artificial Intelligence is contained in millions of huge mathematical matrices. But rather than representing elements of an image, the elements of an AI matrix represent small language segments converted to numbers called "tokens."

Initially, these tokens have no relationship, no understanding of each other. So, the AI system must be trained by showing it examples drawn from literature, books, databases, and myriad other sources. Slowly, the system discerns patterns. How sentences flow. What words form phrases. What words do things. What words are things.

Years before a child studies grammar, through repetition and example she learns words are different serving as verbs, nouns, adjectives... AI learns in a similar manner learning how words flow. For every word, part of a word, or phrase, AI systems are trained by making guesses based on probabilities, learning from the results, and then adjusting those probabilities.

That slowly acquired knowledge is based on increasingly refined guesses after "reading" billions of documents. For example, the word "Address" following "Gettysburg" is most likely referring to President Lincoln's speech but could also be referring to a street address in Gettysburg, Pennsylvania. Based on the probabilities calculated for neighboring words, AI can discern the actual meaning.

Training a general-purpose AI system requires on the order of septillion calculations—that's ten with twenty-four zeroes behind it. If you're wondering how long it would take to count to a septillion, don't. It's far, far longer than the life of the universe as we understand it.

Given the amount of computation required, it's remarkable that we have the technology to train AI systems. But it's massively expensive. General knowledge AI systems require months of training time and as many as 100,000 Nvidia GPUs each making trillions of calculations per second. As ChatGPT described its own training:

> "Training AI is like teaching a child to read and predict language—only the "child" practices billions of times faster, with trillions of examples, using 100,000 supercharged digital tutors working in parallel."

# The Danger of Armed Government Intervention

DECEMBER 2, 2025

Last Wednesday, an Afghan refugee in a targeted attack shot two West Virginia National Guardsmen, Sarah Beckstrom and Andrew Wolfe. The Guardsmen had been on patrol in Washington, D.C. as part of President Trump's crime-fighting mission.

Sarah Beckstrom died the next day, only twenty years old. Andrew Wolfe, twenty-four, remains in serious condition five days after being shot.

The assassin, Rahmanullah Lakanwal, twenty-nine, is an Afghan refugee. He entered the United States in 2021 through Operation Allies Welcome, a program that resettled Afghans after the U.S. withdrawal from the country. Lakanwal had served in a CIA-backed Afghan Army unit for ten years and had been granted political asylum this year in April.

The attack generated a fierce response from the White House. The day after the shooting, President Trump wrote in a TruthSocial post that he would "permanently pause migration from "all Third World Countries," declaring, "This refugee burden is the leading cause of social dysfunction in America." The United States Citizenship and Immigration Services (USCIS) director, Joseph Edlow, said in a statement he was directing a "full-scale, rigorous re-examination of every green card for every alien from every country of concern," at President Trump's request.

President Trump's anger was understandable, but "the refugee burden is [not] the leading cause of social dysfunction in America." The unprovoked attack on the two young Guardsmen could easily have been done by a red-blooded American as happened in the two assassination attempts on Donald Trump and the murder of Charlie Kirk.

Federal troops escorting Black Little Rock High School students in September 1957 and policing the nation's Capital in November 2025.

Thomas Crooks was killed on August 13, 2024, attempting to assassinate Donald Trump during a Pennsylvania campaign rally. A U.S. born citizen of American parents, Crooks was a member of the National Technical Honor Society and graduated from Bethel Park, Pennsylvania High School with high honors.

A month later on September 15, Ryan Wesley Routh was arrested in an attempt to assassinate Trump while playing golf at his West Palm Beach golf course. Routh was a U.S. citizen born of American parents. He once owned a roofing company employing ninety employees. Based on his 2024 mug shot, Routh could be mistaken for an aging, blond surfer dude.

A year later on September 10, 2025, conservative-advocate Charlie Kirk was murdered by Tyler James Robinson. Robinson who born and raised in Washington, Utah. At the time of his arrest, Robinson was in his third year of an electrical apprenticeship at Dixie Technical College. He and his family are members of the Church of Jesus Christ of Latter-day Saints.

Still, President Trump is rightly concerned that 200,000 Afghans have entered the United States since Afghanistan's collapse to the Taliban in August 2021. An unknown number were trained by the CIA as killers to fight alongside U.S. Army troops in the brutal war against the Taliban and ISIS terrorists. These are potentially dangerous people whether due to Post-Traumatic Stress Disorder (PTSD) or other causes.

The Trump administration has a responsibility to assure, as much as possible, that Afghans and other immigrants are not a threat to Americans.

Like today, during the early nineteen-nineties most Americans wanted the government to get tougher fighting crime. But two events convinced a segment of Americans that a tyrannical government was the greater threat.

In August 1992, hundreds of law enforcement and federal agents converged on the Ruby Ridge, Idaho cabin of Randy Weaver. Weaver had repeatedly failed to appear for trial on illegal weapons charges. Weaver and his family were religious fundamentalists who had been peacefully living in their isolated cabin without electricity or running water since 1984. The siege lasted for eleven days including a firefight during which a U.S. marshall was killed. Weaver only surrendered after his wife and son were killed by FBI snipers. Tried for murder and conspiracy, Weaver was only convicted for failing to appear for trial on the original weapons charge.

Six months later, a massive federal siege led to far more deaths. On February 28, 1993, federal agents raided the Waco, Texas compound of David Koresh, the founder and leader of the Branch Davidian Christian sect. The charges, like at Ruby Ridge, were federal firearms violations. There were also rumors that Koresh was sexually abusing young girls. Four federal agents and six Davidians were killed in the raid. A fifty-one-day siege ensued, viewed on television by millions around the world. Koresh eventually announced he and his followers would surrender once he had transcribed a message he believed God had sent him. But the FBI, impatient after 754 phone calls spent negotiating with Koresh, chose to end the siege on April 19 by driving two armored vehicles into the compound. Fires erupted that soon engulfed the building. As the siege ended, seventy-six Davidians, including twenty-five children, were dead, consumed by fire and self-inflicted gunshots. Most Americans viewed these events as human tragedies, but a growing number considered them confirmation of the federal government's tyranny.

The federal government's siege of the Branch Davidian compound in Waco, Texas resulted in 76 deaths, including 25 children.

Two years to the day after federal agents stormed the Branch Davidians' Waco compound, on April 19, 1995, Timothy McVeigh detonated a bomb laden truck in front of the Murrah Federal Building in Oklahoma City. The blast killed 168 people and wounded hundreds more. It remains the greatest act of home-grown terrorism in U.S. history. McVeigh, a decorated veteran of the First Gulf War, had been radicalized by the federal actions at Ruby Ridge and, especially, Waco.

Shortly before his execution, McVeigh sent a letter to Fox News explaining his twisted justification for the bombing. It reads, in part:

> "The bombing was a retaliatory strike; a counterattack, for the cumulative raids (and subsequent violence and damage) that federal agents had participated in over the preceding years (including, but not limited to, Waco.) From the formation of such units as the FBI's 'Hostage Rescue' and other assault teams amongst federal agencies during the '80's; culminating in the Waco incident, federal actions grew increasingly militaristic and violent, to the point where at Waco, our government—like the Chinese—was deploying tanks against its own citizens."

In 1964, based on Gallup polls, 77 percent of Americans trusted their government to do the right thing. That trust began to deteriorate in the nineteen-sixties with the Vietnam War, followed by the Watergate scandal in the nineteen-seventies and the Iran-Contra scandal in the nineteen-eighties. The government sieges of Ruby Ridge and Waco further disillusioned many Americans. By June 1994, public trust in government had fallen to 17 percent, a postwar low. Public trust would never again approach its earlier, far higher levels. (The preceding section was excerpted from *We The Presidents*, pages 428 - 430.)

Broadcast live on television, the Ruby Ridge and Waco sieges fueled a deep distrust of the government. In 1995, Wayne LaPierre of the National Rifle Association (NRA) in a fundraising letter (later retracted) regarding a ban on assault weapons, wrote, "[The gun ban] gives jack-booted government thugs more power to take away our constitutional rights, break in our doors, seize our guns, destroy our property, and even injure or kill us."

Decades later, government distrust remains the *raison d'être* for today's militia groups including the Oath-Keepers, Three-Percenters, and Proud Boys. In 2020, believing the "Deep State" had stolen the election from President Trump, these and other anti-government groups raided the U.S. Capitol—the worst civil insurrection since the Civil War.

Today, heavily armed and masked government troops are patrolling America's cities in search of criminals and illegal immigrants. Hopefully, both soldiers and civilians will exercise restraint. When armed federal enforcers and civilians mix, history shows that simple incidents can quickly escalate, just as happened at Ruby Ridge, Idaho and Waco, Texas.

# Immigration and Violent Crime

DECEMBER 9, 2025

Violent crime—murder, rape, assault, and robbery—has been in the headlines nearly every week since President Trump took office on January 21. The President blames foreign-born immigrants for much of that crime.

So, this should be an easy question: Of the following five U.S. cities, which city had the highest violent crime rate in 2024? Which city had the lowest?

- Detroit, Michigan
- El Paso, Texas
- Nashville, Tennessee
- New York, New York
- San Francisco, California

Reducing violent crime has been one of President Trump's highest priorities. During his inaugural speech on January 20, President Trump promised, "to use the full and immense power of federal and state law enforcement to eliminate the presence of all foreign gangs and criminal networks bringing devastating crime to U.S. soil.... We are going to bring law and order back to our cities."

The President is right that reducing violent crime should be a national priority. Since the end of the Second World War, the United States has suffered the highest murder rate within the twenty largest developed countries.

In 2023, the latest international data available, the U.S. murder rate was 5.9 murders per 100,000 population. The second highest country, Canada, had 2.0 murders per 100,000, a third that of the United States. The next eighteen countries averaged under one murder per 100,000. Japan had a minuscule .23 murders per 100,000. A Japanese city with a population of 500,000 would average slightly over one murder per year. Atlanta, Georgia, with a similar population, had 135 murders in 2023.

El Paso, Texas has a rich heritage of both Mexican and American culture.

But unlike Japan, the United States is hardly a homogenous country. The U.S. population varies widely by race, ethnicity, and citizenship. Today, 58 percent of the U.S. population are non-Hispanic White, 13 percent are non-Hispanic Black, 20 percent are Hispanic (both White and Black), and 6 percent are Asian.

In 2023, approximately 15 percent of the U.S. population, 51.8 million, were foreign-born immigrants including naturalized citizens, lawful permanent residents, temporary migrants, and illegal migrants.

America's illegal immigrant population has skyrocketed, from 3.5 million in 1990 to 14 million in 2023. With such a large increase in illegal immigrants, it seems inevitable that violent crime would have significantly increased since 1990. As Donald Trump has often claimed at his rallies, "Unauthorized immigrants are criminals… snakes that bite… coming from jails and mental institutions… causing crime in sanctuary cities… and are killing Americans en masse."

---

So, which of the five cities did you pick as having the highest violent crime rate? If you had a hard time choosing between Detroit, San Francisco, and New York, that's not surprising. These cities are often accused of being "under invasion from within" by foreign-born immigrants. Yet, although Detroit has a relatively small immigrant population, it had the highest violent crime rate, 35 murders per 100,000 population. Perhaps more surprising, San Francisco and New York, with large immigrant populations, had murder rates below the national average.

Below are the 2024 actual crime statistics based on FBI violent crime data.

1. Detroit, Michigan ( 35.0 murders per 100,000 | 9.5% foreign-born immigrants)
2. Nashville, Tennessee ( 14.5 | 15.9% )
3. San Francisco, California ( 5.2 | 34.9% )
4. New York, New York ( 4.7 | 36.3% )
5. El Paso, Texas ( 2.3 | 26.9% )

Contrary to claims by our President and Vice-President, cities with high immigrant populations tend to have low violent crime rates. San Francisco and New York are depicted as being crime-ridden; they certainly have issues with homelessness and related crimes. But both cities, with large immigrant populations, have violent crime rates below the national average.

And what about El Paso just across the Rio Grande from Juárez, Mexico? More than a quarter of El Paso's population is foreign-born, yet El Paso has the lowest murder rate and third lowest violent crime rate of the nation's twenty-five largest cities.

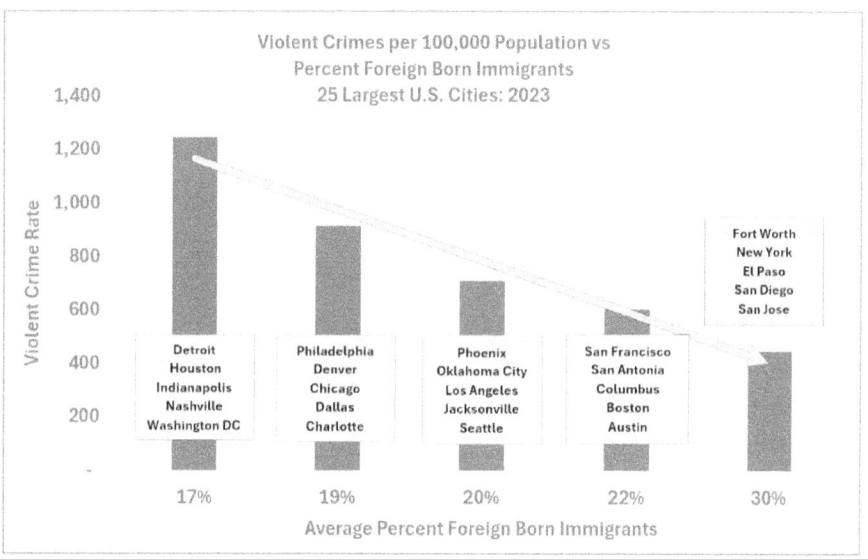

Cities with large foreign-born immigrant populations typically have low violent crime rates. FBI

The reality is that the vast majority of immigrants, both legal and illegal, just want a better life for themselves and their families. Millions of immigrants work quietly in construction, landscaping, agriculture, hotels, restaurants, home cleaning, and other low-paying, labor-intensive jobs. Jobs that others shun.

A small percentage of immigrants though are truly evil such as MS-13 gang members, members of drug cartels, and murderers like Jose Antonio Ibarra—an illegal immigrant who brutally murdered 22-year-old nursing student Laken Riley while she was jogging. But study after study have concluded that, on whole, immigrants are less likely to commit violent crimes than the general population. Here's why, as one study by the non-partisan Cato Institute concluded:

> "Undocumented migrants commit fewer homicides for many reasons. First, the punishments are harsher – they get deported. Second, many came from more violent countries because they wanted more safety. Third, they mostly leave their families, friends and cultures behind because they want a better future for themselves and their children. People like that are just less likely to be criminals in the first place."

In short, reducing the immigrant population is not the simple solution to decreasing America's violent crimes many claim. Furthermore, sending in combat-ready troops to patrol streets armed with assault weapons may temporarily depress crime, but unless the root causes are corrected crime quickly returns when the troops leave.

Reducing violent crime is hard work requiring planning, persistence, and police. In this regard, few cities have done a better job than New York City. In 1990, New York was plagued by rampant violence, drug wars, and urban decay. That year, the City suffered 2,245 murders—a murder, on average, every four hours. By 2001, murders had fallen to 649, a 71 percent reduction, a reduction that significantly exceeded national trends.

Rudy Giuliani, the City's mayor from 1994 through 2001, deserves much of the credit. Shortly after becoming mayor, Giuliani appointed William Bratton as Police Commissioner. It was an excellent choice. Although Bratton only served for two years under Mayor Giuliani, he transformed the New York Police Department.

Bratton started by cracking down on minor offences such as subway graffiti, fare evasion, broken windows, and public disorder. This strategy sent a strong message that the City was serious about reducing crime, even minor crimes.

Bratton next installed CompStat (Computer Statistics), a data-driven system that had quietly been developed by a New York Transit detective for quickly pin-pointing crime. Compstat mapped crime hotspots in real-time, allowing rapid police intervention. To increase accountability, during weekly meetings Police Commanders were required to report crime trends in their precincts. CompStat shifted the city's policing from reactive to proactive, using data to anticipate and prevent crime rather than responding to incidents after they occurred.

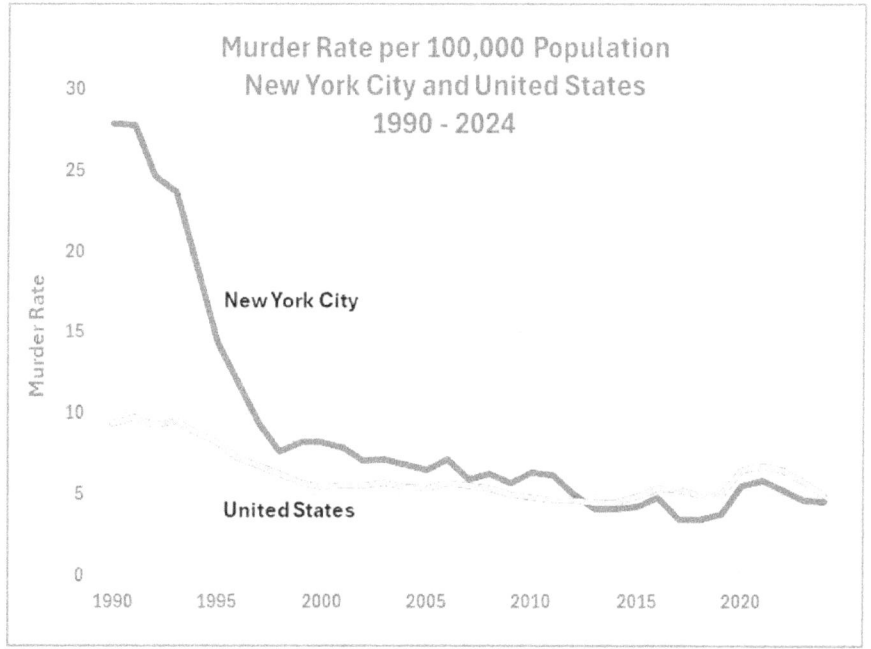

Once one of the highest in the nation, New York City's murder rate today is below the national average. FBI

These two initiatives were supported by a large increase in the city's police force from 26,000 in 1990 to nearly 40,000 by the late nineteen-nineties. The increase allowed more patrols in high crime areas as well as gun interdictions including "stop and frisk" which allowed police officers to stop suspicious individuals to check for illegal weapons.

Many of the new policies were controversial at the time. But they worked. By 2001, New York City had become the safest large city in America—even while its foreign-born immigrant population increased nearly 50 percent from 2.0 million in 1990 to 2.9 million in 2001.

Reducing crime is hard work on multiple fronts. New York has shown the way. Turning the United States into a police state with heavily armed, masked soldiers patrolling American streets is not the answer.

# Money and the Media

DECEMBER 16, 2025

By 1952, Willie Sutton had robbed nearly 100 banks and made three prison escapes during his three decades as a bank robber. Unlike John Dillinger, Baby Face Nelson, and Pretty Boy Floyd, Willie Sutton never killed anyone; often his gun wasn't even loaded. Unfailingly polite, Sutton treated bank employees and customers with care, even abandoning a robbery if a child or woman cried. According to the FBI, one victim said witnessing one of Sutton's robberies was "like being at the movies, except the usher had a gun."

More than as a bank robber, Willie Sutton became famous for being honest. Shortly after his arrest in February 1952, a reporter asked him why he robbed banks. "Because that's where the money is," Sutton replied. Sutton's disarming honesty instantly made him an American folk hero.

Years later, in his autobiography, *That's Where the Money Was*, Sutton wrote he never actually said he robbed banks "Because that's where the money is." Instead, he wrote that he robbed banks, "Because I enjoyed it. I loved it. I was more alive when I was inside a bank, robbing it, than at any other time in my life."

In either case, Willie Sutton was disarmingly honest.

On December 4th, Americans witnessed another case of disarming honesty.

That day, FBI Deputy Director Dan Bongino, with Attorney General Pam Bondi and FBI Director Kash Patel, announced the FBI had arrested Brian J. Cole Jr., the man they believed responsible for planting pipe bombs near Republican and Democratic party headquarters on the eve of the January 6, 2021, Capitol Insurrection.

Dan Bongino as a prominent podcaster and later being sworn in as the FBI Deputy Director on March 17, 2025.  WIKIPEDIA | FBI

During the press conference, Bongino praised FBI agents for solving the case based on re-examination of existing forensic evidence. Bongino was effusive in his praise for President Trump and Attorney General Pam Bondi, but also praised FBI agents for "tracking this person to the end of the earth. There was no way he was getting away…"

For millions, Bongino's praise of the FBI was shocking. Before his appointment as FBI Deputy Director in March 2025, Bongino had been a hugely popular podcaster known for promoting Deep State conspiracy theories, many regarding the FBI.

Bongino's best-selling book, *Follow the Money*, claimed to expose "the labyrinth of connections between D.C.'s slimiest swamp creatures - Democrat operatives, lying informants, desperate, and destructive FBI agents…" After the January 6, 2021, Capitol Insurrection, Bongino persistently claimed the "FBI knows the identity of this pipe bomber…Folks, this guy was an insider. This was an inside job. And it is the biggest scandal in FBI history."

But during the FBI press conference, Bongino never mentioned his prior claim that the FBI was "irredeemably corrupt," or that the FBI should be disbanded and its agents fired. That's not surprising. According to the *Wall Street Journal*, Deputy Director Bongino never attempted to uncover the alleged Deep State within the FBI. Rather, "In his first days as deputy director in March, Borgino ordered a briefing of the bombs and declared solving the case one of his priorities."

The breakthrough came when a tech-savvy agent was finally able to read a tranche of encrypted cell phone data provided by T-Mobile. Once Brian J. Cole Jr. was arrested, Cole, rather being an FBI insider, "expressed support for Trump and said he had embraced conspiracy theories regarding Trump's 2020 election loss."

A few hours after the FBI press conference, Bongino appeared on Fox News' Sean Hannity Show. Not surprisingly, Hannity reminded Bongino:

> "You said there's a massive cover-up because the person who planted those pipe bombs, they don't want you to know who it is because it's either a connected anti-Trump insider or an inside job. You said that, you know, long before you were even thought of as deputy FBI director."

Rather than defend, or even mention, his earlier conspiracy claims, Bongino was honest and simply replied:

> "Listen, I was paid in the past, Sean, for my opinions. That's clear. And one day I will be back in that space, but that's not what I'm paid for now. I'm paid to be your deputy director, and we base investigations on facts."

Bongino's comment was widely interpreted as walking back his earlier claims that the FBI, as part of the Deep State, was involved in the pipe bombs and subsequent cover-up.

I admire Bongino's candor. Bongino's comments that day—both during the FBI press conference and the Sean Hannity Show—showed honesty and respect for his role as the FBI Deputy Director. As he told Hannity, his FBI job is based on facts, not opinion.

But that wasn't the case in his prior job as a podcaster. As a podcaster, Bongino understood the rules separating facts and opinion were different. In today's media world, often there is little separation.

As a former NYPD police officer and Secret Service agent, Bongino had the credibility to make whatever claims he wished as a podcaster. He chose to focus on "a vast, well-funded cabal of wealthy Democrats and D.C. swamp elite…" The strategy was wildly successful. By 2024, Bongino had built one of the largest audiences in social media with hundreds of millions of downloads a month. Earlier this year, *Forbes* magazine estimated Bongino's fortune to be over $150 million.

(Note: Monday evening Fox News reported that "Deputy FBI Director Dan Bongino to decide about future at bureau in coming weeks, sources say.")

Today, much of media is driven by alternative facts, conspiracy theories, and partisan division. Above all, the objective is to build a large, loyal, and lucrative audience.

We saw that when Infowars claimed the 2012 Sandy Hook school murders were a staged event, with the Steele Dossier and Hunter Biden's notebook scandals promoted by "mainstream media," with Robert F. Kennedy Jr's claim during the Covid pandemic that vaccines killed more people than Covid itself, and perhaps the greatest false claim of all, the claim that corrupted voting machines had stolen the 2020 presidential election from Donald Trump.

No media promoted the stolen election claim more than Fox News.

Fox began election night on November 3, 2020, as on honest news broker when their election desk called a Biden win for Arizona. President Trump's supporters were furious. Many switched channels to Newsmax, Fox's scrappy, new competitor.

To placate its rebellious viewers, by November 5 Fox was beginning to report "voter fraud at a frightening level... [and] accounts from Americans who witnessed voter fraud firsthand." Three days later, Fox began promoting claims that Dominion Voting Systems, somehow supported by Venezuela, had rigged the 2020 election in favor of Joe Biden. For months, Fox pundits including Lou Dobbs, Jeanine Pirro, Maria Bartiromo, Tucker Carlson, and Sean Hannity promoted the election fraud claims during unchallenged interviews with conspiracists Sydney Powell, Rudy Guiliani, Mike Lindell, and others.

But Fox, its management and hosts, never believed the election fraud claims. Less than two weeks after the election, Fox host Laura Ingraham emailed Tucker Carlson and Sean Hannity that "Sidney Powell is a bit nuts." Tucker Carlson was more candid, emailing his producer, "Sidney Powell is lying." "That whole narrative that Sidney [Powell] was pushing, I did not believe it for one second," Hannity later testified in a deposition. Regarding the MyPillow CEO, Mike Lindell, Fox employees emailed among themselves that Lindell has "really gone off the deep end" and is "definitely crazy." In emails and texts, Fox chairman Rupert Murdoch referred to Giuliani's statements as "crazy stuff" and called Sydney Powell a "crazy, would-be lawyer."

Yet, Fox News for months promoted claims that Dominion voting machines had been rigged to swing the election to Joe Biden. Falsely accused of election manipulation, Dominion Voting Systems sued Fox News for defamation in March 2021.

Caught promoting election fraud claims they privately mocked, Fox News management settled the Dominion defamation lawsuit in April 2023 for $787.5 million—the largest defamation settlement in U.S. history.

*Fox* never apologized to its viewers for its false claims, instead issuing a short statement declaring, "We acknowledge the Court's rulings finding certain claims about Dominion to be false."

# Letters from the Past: Four Presidents on Today's America

DECEMBER 23, 2025

As the year 2025 comes to a close, I asked four of our greatest presidents—George Washington, Abraham Lincoln, Franklin Roosevelt, and Ronald Reagan—for their thoughts on the state of America today. Artificial Intelligence was, of course, involved. Yes, I know. Many of us are tired of AI, all day, every day. But, bear with me.

For each president, starting with George Washington, I made the following request to ChatGPT, Claude, Gemini, and Grok:

> I have a three part request: (1) Study the writings and speeches of George Washington, (2) read the major articles and editorials in the Wall Street Journal and New York Times published in 2025, and (3) assume you are George Washington and write a 250 word letter to the citizens of the United States offering your opinion as to the state of America today.

I chose the conservative *Wall Street Journal* and liberal *New York Times* as they are, in my opinion, responsible journalists. The results from the four AI systems were surprisingly similar—with a few exceptions. Below are the four letters generated by Google Gemini, the AI system I arbitrarily selected. It's easy to ask your preferred AI the same questions.

SHORT ESSAYS FOR INQUIRING MINDS

Take a few minutes to read the heartfelt letters that AI suggests Washington, Lincoln, Roosevelt, and Reagan would write to Americans were they alive today.

---

## George Washington

To the Friends and Fellow-Citizens of the United States,

December 23, 2025

It is with a heavy heart that I observe the condition of our beloved Union from the vantage of history. When I bid you farewell in 1796, I warned that the "spirit of party" was the worst enemy of popular government. Today, I see this spirit kindled into a consuming fire. Your nation is riven by factions that seem less like fellow citizens and more like hostile nations, each seeking to ruin the other rather than advance the common good.

I perceive that the "baneful effects" of this partisanship have opened the door to foreign intrigue and domestic instability. Your public debt, now grown to a ruinous magnitude, ungenerously throws upon posterity the burden which you yourselves ought to bear. I see your commerce entangled in foreign disputes and your peace threatened by antipathies against some nations and passionate attachments to others—precisely the "foreign entanglements" I implored you to avoid.

The preservation of your liberty depends on the preservation of your Union. You must frown upon the first dawning of every attempt to alienate any portion of our country from the rest. Do not let the cunning of ambitious men, who seek to aggrandize themselves on the ruins of public liberty, divide you.

I beseech you: restore your national morality, cherish public credit, and above all, subordinate your local and party prejudices to the immense value of your Union. Without it, your independence is but a name.

Your humble servant,

G. Washington

---

## Abraham Lincoln

To the Citizens of these United States,

December 23, 2025

I address you not as a partisan, but as a humble servant of the Union whose preservation was my life's greatest burden. I have looked upon your newspapers and observed the temper of your times. It grieves me to see a house once again so perilously divided against itself.

In my own time, I warned that "passion has helped us; but can do so no more." Today, I see a nation intoxicated by the spirit of party and the thirst for retribution. I read of leaders described as possessing an "alcoholic's personality"—driven by impulse and chaos—and a populace that seems to crave the breakage of norms as a drunkard craves the bottle. But I say to you: Reason—cold, calculating, unimpassioned reason—must be the pill you swallow if you are to survive.

You are debating tariffs and tribunals, borders and banishments. These are grave matters. Yet, I fear you are forgetting that the Constitution is not a weapon to be wielded against your neighbors, but a covenant that binds you to them. If you allow the "Rule of Law" to become merely the "Rule of Will," you tear at the mystic chords of memory that hold this vast Republic together.

We must not be enemies. Though passion may have strained, it must not break our bonds of affection. I beseech you to turn from the intoxicating rhetoric of division and listen once more to the better angels of your nature.

Yours in the bond of Union,

A. Lincoln

---

## Franklin Roosevelt

To the American People,

December 23, 2025

My Friends,

I speak to you across the span of history, looking upon a nation that faces a "rendezvous with destiny" as critical as any in my own time. I read your journals and see a people gripped by uncertainty, where the "economic royalists" of the digital age have amassed power that rivals the industrial monopolies I once fought.

In 1933, I told you that the only thing we had to fear was fear itself—nameless, unreasoning, unjustified terror which paralyzes needed efforts. Today, I fear that terror is being manufactured and sold to you for political profit. I see a nation tempted to trade its essential liberties for the illusion of security, and to mistake the bluster of autocracy for the strength of leadership.

You are debating the dismantling of the social contract we forged in the fires of the Great Depression. Remember that a necessitous man is not a free man. If you allow the dismantling of your safeguards—your security in old age, your protection from the excesses of the market—you surrender the very foundation of your freedom.

I see, too, a dangerous retreat into isolationism. We learned, at the cost of much blood, that we cannot live as an island. To turn inward, to view your neighbors as enemies and your allies as burdens, is to invite the very chaos you seek to avoid.

Do not lose faith in democracy. It is not a static thing; it is an unending labor. Reject the politics of resentment. Reclaim your confidence in one another.

Faithfully,

Franklin D. Roosevelt

---

## Ronald Reagan

My Fellow Americans,

December 23, 2025

I once spoke of our nation as a shining city on a hill—teeming with people of all kinds living in harmony and peace. Looking at you now, I wonder if we have let the lights dim, not because of enemies without, but because of anger within.

I read your headlines and see a people debating the value of walls rather than bridges and turning inward when the world so desperately needs American strength. I always believed that if we unleashed the creative energy of the individual, there were no limits to our growth. But today, I fear you are trading the open seas of free commerce for the stagnant waters of protectionism, forgetting that a rising tide lifts all boats.

Freedom is a fragile thing, and it is never more than one generation away from extinction. It is not ours by inheritance; it must be fought for and defended. When I see Americans questioning the very institutions of our democracy, or looking to a strongman to solve problems that only a free people can solve, I worry you are forgetting who you are. Government is not the solution to your problem; government is the problem when it seeks to dictate your conscience or your commerce.

We must remember that our opponents are not our enemies. I fought hard against Tip O'Neill during the day, but we were friends after 6 p.m. You must rediscover that grace. Do not be afraid of the future. You are Americans. There is nothing you cannot do if you do it together.

God bless you, and God bless the United States of America.

Ronald Reagan

---

*Today's Substack was inspired by David Parson's latest book, Regulation: Freedom within Limits The book documents an AI-based conversation among experts regarding the effects of regulation on individual and community rights.*

# The Short, Miserable Life of a Golf Ball

DECEMBER 30, 2025

Today is my last Substack post for 2025. I've written fifty-one posts this past year ranging from Cracker Barrel's logo to the Ukraine War. So, to close out the year on a truly worthy topic, let me ask: Have you ever considered the short, miserable life of a golf ball?

SHUTTERSTOCK

Starting with the early Greek philosophers, a fantastic theory, panpsychism, suggests that all matter, from rocks to humans, are conscience. For sure, nobody is claiming that rocks worry about the federal debt. Still, maybe the guy who invented Pet Rocks back in the seventies was on to something.

Actually, I don't think rocks and other inanimate objects, can think. But what if they could? What might a golf ball think? Who knows, but we can imagine:

"I was born in a cardboard sleeve, one of three identical triplets, pristine white with perfect dimples. My siblings and I each had our own dreams: wide fairways, straight flights, gentle landings. Maybe even a PGA tour followed by retirement on a trophy shelf.

"But there's no PGA pro.

"Only the Giant. His gloved hand rudely grabs me and places me upon a wooden stake shoved into the hard earth. For a brief moment, I see everything: a beautiful, green fairway; swaying, majestic trees to the right; blue water to the left; a rippling flag in the distance.

"But now the Giant's shadow looms. I feel the whoosh of practice swings, each one a near-miss that rattles me to the core. Finally, the downswing. A massive clubface hurtles toward me at 120 miles per hour.

"CRACK!

"In a fraction of a second, I am crushed flat with a force of 2,000 pounds, my rubber heart compressed in agony. I rebound, spinning wildly at 3,000 RPM, launched into the sky. For a brief moment, I feel sublime exhilaration—soaring high, wind rushing over my dimples, defying gravity.

"But it's fleeting. The Giant's swing was wild. I am heading—not for the wide, green fairway—but for the water. I land with a plop and sink slowly into the murky depths where I join my battered brethren on the dark, muddy bottom.

"My only hope is to be retrieved by a diver and later sold as a "water ball." It's an ignoble ending, destined to be hit into the dirt, the sand hazards, the tall grass, and finally, irretrievably lost whether to the water or the woods.

"I started life with so many dreams. But today, I am lost and forever forgotten. Such is the short, miserable life of a golf ball."

So, if you're a golfer, the next time you hit a bad shot, just think how the poor golf ball feels.

---

As an engineer, I'm tempted to finish today's short post with a discussion on the physics of golf which are quite interesting. I'll spare you that, except for two surprising facts.

1. A professional golfer's 300-yard drive is in the air for roughly six seconds—time for a light beam to travel 1.1 million miles, or two round trips to the moon. But during that brief instant when the driver struck the ball, the club and ball remained in contact for a mere 500 millionths of a second; only time for light to travel 75 miles, not even the distance from Naples to Miami.

2. The contact between the driver and golf ball may be vanishingly brief, but it is one of the most violent events in sports. During that short interaction of club and ball, approximately 150 Joules of energy are released. An electrical generator producing that much energy continuously would generate 300 kilowatts, easily enough to power twenty-five average-sized homes.

Maybe sometime I'll write about the easy life of a baseball. No cold water or dark woods for them. Hit into the stands, baseballs are taken home and treasured by small boys.

# SHORT ESSAYS FOR INQUIRING MINDS

# About the Author

Ronald Gruner founded and served as the chief executive of three successful technology firms over his forty-year career. As a business leader, Gruner wrote extensively on industry and technology. Since retiring, Gruner has published two award-winning histories: *We The Presidents* and *Covid Wars*. Gruner lives with his wife, Nancy, in Naples, Florida where he is a consistently poor golfer.

If you enjoyed *Short Essays for Inquiring Minds*, you may also enjoy Gruner's weekly Substack postings on current events, history, science, and other topics of interest to inquiring minds.

RonaldGruner.substack.com

SHORT ESSAYS FOR INQUIRING MINDS

# Other Books by Ronald Gruner

"So, you think you know your presidential history? Think again."

Jack Falvey,
Opinion Writer
*The Wall Street Journal*

"A Covid-19 reference as comprehensive as it is devastating."

*Kirkus Reviews*

# Index

## A

Abbott, Greg, 281
abortion, 171, 251, 258
Abraham Accords, 169
Abrams, Stacey, 282
Adams, Abigail, 60
Adams, Eric, 239
Adams, John, 60
Afghanistan: refugees from, 306, 308; Taliban's rule in, 308; US withdrawal from, 171
Aircraft Owners and Pilots Association (AOPA), 294
Air Mail Act of 1925, 56
air traffic control (ATC) systems: controllers' work schedule, 292–93, 294; establishment of, 289–90; international, 293–94; privatization of, 294; public safety and, 288, 293; "shrimp boat" markers in, 290; technology advancements and, 290–91, 292
Alexander II, Emperor of Russia, 10
Alexander III, Emperor of Russia, 3
AlexNet, 303
alternative facts, 45, 278, 322
Amazon, 301
American Anti-Vaccination Society, 62
American Civil War, 97, 255–56, 311
American Exceptionalism, 71

American Institute for Economic Research, 64
American Rescue Plan, 170
Americans for Tax Reform (ATR), 216
American War for Independence, 59
America's Frontline Doctors (AFD), 42, 43
Anglo-Persian Oil Company, 160, 161, 162
animals: consumption of wild, 17; disease transmission and, 16–18
Anti-Deficiency Act (1884), 259
Anti-Vaccination League, 61
anti-vaccination movement, 60, 61–63, 68, 236–37, 322
anti-woke movement, 230–31, 233–34, 247, 249
Apple iPhone, 232
Arabian-American Oil Company (Aramco), 160
Argentina: economic development of, 283; fiscal crisis in, 270; Milei's presidency in, 284–86; military dictatorships in, 283–84; US aid to, 283, 284, 286–87
Argo (covert operation), 163
Arizona: COVID-19 pandemic in, 34; shootings in, 245
Arkansas: COVID-19 pandemic in, 33–34; schools in, 82
Artificial Intelligence (AI): datasets, 191; Large

Language Model of, 301;
news making, 277, 278–79;
potential for intelligent
behavior, 180, 275–76, 301;
presidential addresses
generated by, 325–30;
summary of "No Kings"
protests, 279–82;
technological development
of, 186–92, 302–4; training
process, 192, 275, 304–5; use
of language segments
(tokens), 275, 304–5
artificial neural networks, 189,
191–92, 192
Asian flu pandemic of 1957-1958,
23
Atlas, Scott W., 35, 48–49, 51,
64, 65
atomic bomb, 299
Aultman Company, 99
Aunt Jemima, 231
Australia: ATC system in, 293;
educational system in, 84
autism, 62
aviation accidents, 118, 288, 291,
291–92

# B

bacteria, 3
Bakker, James, 22
bank panic, 101
Bannon, Steve, 127
Barry, John: The Great
Influenza, 9, 10
Bartiromo, Maria, 323
Baum, L. Frank, 108, 109
Bayes, Thomas, 271–73
Bayes Theorem, 273, 274–75
Beckstrom, Sarah, 306
Beijerinck, Martinus, 4–5
Bell, Alexander Graham, 298
Berlin Airlift, 211–15
Berlin Wall, 211
Berra, Yogi, 144
Bhattacharya, Jay, 64, 65
Biden, Hunter, 322
Biden, Joe: 2020 presidential
election, 69–70, 323;
ChatGPT analysis of
presidency of, 167–68, 170–71, 172;
ChatGPT psychological
profile of, 152–54, 155–57;
communication style of, 153;
COVID-19 management, 170;
economic policies of, 171;
Emmett Till Antilynching
Act, 55; environmental policy
of, 119; foreign policy of, 153,
171; inauguration of, 118; as
institutionalist, 158–59, 168,
170–71; leadership style of,
152–53, 172; legacy of, 156;
official portrait of, 153;
public engagements of, 153–54, 156
Biglari, Sardar, 230
Bill and Melinda Gates
Foundation, 42
Bill of Rights, 224
Binghamton Immigration Center
Shooting, 245
Bipartisan Policy Center, 260
birthright citizenship, 295–96
Biscuits n' Gravy, 233
Black Americans: civil rights of,
54–55, 243, 307; violence
against, 243–46
Bloch, Felix, 299
Blockbuster, 233
Bluecoats Drum and Bugle
Corps, 255
Blumenthal, Paul, 122
Bolsonaro, Jair, 206
Bondi, Pam, 320
Bongino, Dan, 320, 320–22;
Follow the Money, 320
Boole, George, 181–82, 183; "An
Investigation of the Laws of
Thought," 181
Borders (retailer), 233
Boston Bruins, 254
Boston Celtics, 254
Boston Consulting Group (BCG),
231

Boston Crusaders, 254–55, 255, 256–57
Boston Patriots, 254
Boston Red Sox, 254
Branch Davidian Christian sect, 308–9, 309
Bratton, William, 316–17
Brazil: STEM graduates in, 298, 299
Breitbart News Network: on Trump's first 100 days in office, 115, 116–17
Brewster, Willie, 244
Brin, Sergey, 299
Brinkley, David, 277
Britton, Elizabeth Ann, 58
Britton, Nan: The President's Daughter, 58
Brown, Chester T., 31
Bryan, William Jennings, 101, 102, 107–8
budget deficits, 138–39
Bud Light, 233, 234
Buenos Aires Times, 285
Bureau of Labor Statistics (BLS), 207, 208, 209, 210
Burroughs Corporation, 93
Bush, George H. W., 26–27, 217
Bush, George W.: collapse of Soviet Bloc and, 78; education policy, 80–81; flu pandemic management, 11–13; legacy of, 26; NATO and, 76; summer reading list, 9; taxation policy, 144; vaccine manufacturers and, 11–12
Bush, Laura, 80–81
Business Advisory Council, 149
business investments, 138

C

California: COVID-19 management in, 66; gerrymandering in, 222, 227; schools in, 82
Camp, Garrett, 130, 131
Canada: ATC system in, 293–94; education system in, 84; homicide rate in, 313
Carbolic Smoke Ball, 40
Carlson, Tucker, 22–23, 110, 323
Carnegie, Andrew, 100
cars: domestic and imported brands, 89–90; gasoline mileage standards, 89
Carter, Jimmy: education policy of, 79–80; "funding gaps" issue, 258; on Great Depression, 147–48; Iran hostage crisis and, 163
La Casta (The Political Caste), 285
"Cathedral of Greenland," 200
Centers for Disease Control and Prevention (CDC), 19, 36, 236, 237
Central Intelligence Agency (CIA): Afghan operations, 306, 308; global elite in, 42, 110, 236; Iran hostage crisis and, 163; reorganization of, 162
Central Processing Unit (CPU), 184–85
Chamberlain, Chuck, 50
Chamberlain, Neville, 75
Chamber of Commerce, 149
Chancellor, John, 277
Charleston church shooting (2015), 246
ChatGPT, 325; analysis of Biden's presidency, 152–54, 155–57, 158–59; analysis of Trump's presidency, 152, 154–55, 157–58, 159; description of AI training, 305; infrastructure of, 301; summary of "No Kings" protests, 279–80
Chevy models, 89
Chicago Cubs, 254
Chicago strike of 1886, 126
Children's Health Defense, 41
China: consumption of wild animals in, 17; COVID-19 outbreak, 16, 19; defence

spendings, 268; interest in Greenland, 194–95; Korean War and, 28; military power of, 268; nationalism, 242, 250–51; STEM graduates in, 298, 299; US tariffs on, 169
CHIPS and Science Act, 170
Churchill, Winston, 50, 160, 161, 273
Church of Jesus Christ of Latter-day Saints, 308
Civil Aeronautics Administration, 292
Civiletti, Benjamin, 258–59
Civil Rights Act (1964), 243
civil rights movement, 123, 242, 243–44
Claude large language model, 279, 280–81, 301, 325
Cleveland airport, 290
Clinton, Bill: ATC system and, 294; budget balancing efforts, 142, 143–44, 262; economic policy of, 81, 140–42; foreign policy of, 287; NATO and, 73, 76, 78; photos of, 74, 141; taxation policy of, 141–42, 143–44
CNN, 23, 94, 120–21
Coca-Cola Company, 234
Coinage Act, 107
Cold War, 214–15
Cole, Brian J., Jr., 320, 321
Colorado: COVID-19 management in, 66
Committee on Economic Security, 149
communism, 27, 31–32
CompStat (Computer Statistics), 317
computers: ability to think, 186, 276; diagram of, 184; hardware, 184–85; logic functions in, 182–84
Compute Unified Device Architecture (CUDA), 303
Congressional Budget Office (CBO), 261–62, 269
Congressional Select Committee on the Coronavirus Pandemic, 111–12, 114
Conservative Political Action Conference (CPAC), 285
Conway, Kellyanne, 45
Coolidge, Calvin, 54, 56, 142, 147, 262, 296, 296–97
corporations: income taxes, 137; pro-profit philosophy of, 92–94, 96, 132
COVID-19 pandemic: conspiracy theories about, 21, 41–42, 44–45, 322; economic impact of, 20, 38, 71, 269; "Focused Protection" approach, 64–67; fraudulent and unproved drugs, 21–22, 43–44; herd immunity, 48–49, 64; lockdowns, 24–25, 33, 34, 81–82; media coverage of, 21, 22–23, 37–38; social distancing during, 49–51; spread of, 19–20, 24, 24; testing guidance, 110; Trump administration response to, 20, 23–25, 33–34, 47, 71, 170; vaccination, 12, 67–68, 110; virus, 15–17, 16, 18, 19, 38, 111, 112, 113
COVID deaths: by age groups, 35–36, 66, 66–67; comorbidities, 37, 37; by education level, 82–83, 83; Great Barrington Declaration and, 64–65, 66–67; international statistics of, 38; in open and close states, 65–66, 66; political orientation and, 69, 69–70; by states, 50, 51, 65–66, 66; statistics of, 24, 25, 33; vaccine hesitancy and, 69
covid.gov website, 110, 111
Covid Wars (Gruner), 72, 82
Cox, James, 53
Cracker Barrel restaurant chain,

229–31, 232–33, 234, 235
crime: foreign-born population and, 312, 314–16, 315; illegal immigrants and, 315–16; statistics of, 312–13
Cronkite, Walter, 277, 278
Crooks, Thomas, 307
Cruz, Ted, 282
Cutter Incident, 62–63
Czech Republic, 76
Czolgosz, Leon, 103

# D

Dahmer, Vernon, 244
Dairy Queen, 233
Daniels, Jonathan, 244
Daugherty, Harry, 52, 57
Deep State, 248, 285, 311, 320–21, 322
Delaware: COVID-19 management in, 66
Democratic Party, 67, 225, 259, 294
De Morgan, Augustus, 181
Denmark: colonies of, 196
Denslow, William, 108
Department of Education Act (1979), 80
Department of Government Efficiency (DOGE), 91, 119, 261, 264
Department of Health and Human Services, 12
Department of War, 240–41
Detroit, MI: crime statistics, 314, 315; foreign-born population in, 315
Dillinger, John, 318
diseases: animals and transmission of, 16–18; caused by viruses, 3–4, 5–6, 7–8; germ theory of, 3, 39
Disney Company, 234
Diversity, Equity, and Inclusion (DEI) policies, 81, 233, 234, 252
Dobbs, Lou, 94, 323

Dominion Voting Systems, 323–24
Doom (video game), 302
Dossier, Steele, 322
Dow Jones Industrial Average, 21, 147
Doyle, Authur Conan, 182
Drum Corps International (DCI), 254, 255, 257
Duke, Mike, 133
Dunkin' Donuts, 233
Durkin, Martin, 26
Dynamic Random Access Memory (DRAM), 183–84

# E

East Berlin, 211, 213
Edison, Thomas, 39
Edlow, Joseph, 306
Edsel Ford, 234
Egede, Hans, 196
Einstein, Albert, 123, 299
Eisenhower, Dwight D.: cabinet of, 26; foreign policy of, 136, 161–62; immigration policy of, 28–29; Interstate Highway System, 29–30; Korean War and, 28; military career of, 136; polio vaccination program, 29; position on taxes, 136; view of leadership, 32
electricity, 39–40, 298, 301, 333
Elizabeth I, Queen of Great Britain, 9
El Paso, TX: crime statistics, 314–15, 315; foreign-born population in, 314, 315, 315
Emergency Quota Act (1921), 54, 296
Emmett Till Antilynching Act (2022), 55
Employment Situation Summary, 238
"End Poverty in California" (EPIC), 148
Enigma codes, 273
Eric the Red, 196

Estes, E.M., 89
Event 201, 42
Evins, Dan, 229
Experimental Aircraft Association (EAA), 294

## F

Facebook, 41, 131, 233
Fairness Doctrine, 277–78
Fall, Albert, 57
Fauci, Anthony, 33, 38, 46–48, 47, 237
Federal Aviation Agency (FAA), 292, 293
Federal Bureau of Investigation (FBI), 320–22
Federal Communications Commission, 56
federal deficit: Congressional responsibility for, 263, 269, 270; defence spendings and, 260, 262, 267–68; management of, 136–37, 138, 140, 259–60, 261–63, 270; Social Security programs and, 264, 270
federal government shutdown, 258
Federal Reserve (FED), 170, 209, 238, 264
Federal Trade Commission (FTC), 21–22
Federal Troops: attacks on, 306, 309; escorting Black Little Rock High School students, 307; operation against civilians, 308–9, 310, 311, 318; operations against illegal immigrants, 311; policing of Capitol, 307
federal workforce, 258, 261–62
Fermi, Enrico, 299
Financial Crisis of 2008, 26
Financial Panic of 1907, 263
Floyd, Charles Arthur "Pretty Boy," 318
Forbes, 322
Forbes, Charles, 57
Ford, Henry, 206
Fort Detrick, 112
Fort Hood Shooting, 245
The Fountainhead (film), 175, 176, 176
Fox News: Bongino interview on, 322; coverage of COVID-19 pandemic, 21, 22–23; on Cracker Barrel's logo change, 230; Dominion Voting Systems lawsuit against, 323–24; McVeigh's letter to, 310; promotion of stolen election claim, 323–24; on Trump's first 100 days in office, 118
France: COVID-19 pandemic, 20; defense spending, 268; economic development of, 283; education system, 84; GDP, 106
Francis, Pope, 12, 124
Franklin, Benjamin, 34–35, 61
fraudulent cures, 21–22, 40, 40–41
Friday's Job Report, 238, 239
Friedman, Milton, 92–94, 93, 96, 132, 133; Capitalism and Freedom, 92
Friendly's restaurant, 233
Fudenberg, John, 37
Fujimori, Alberto, 124
Fuller, Ida May, 145
Full Retirement Age (FRA), 267

## G

gain-of-function research, 113, 114
Gandhi, Mahatma, 244
Garfield, James A., 97, 243
Gates, Bill, 41
General Electric Company, 94
General Motors Company, 26
George VI, King of United Kingdom, 273
Germany: ATC system in, 293; defense spending, 268; deindustrialization of, 211–

12; economic development in, 283; education system in, 84; GDP, 106; reunification of, 76
Gerry, Elbridge, 223, 225
gerrymandering, 223, 225–28, 227
Giffords, Gabby, 245–46
gig economy, 130
Gilbertson, Nick, 116
Gingrich, Newt, 140, 141, 262
global elite, 41–42, 110, 236
Goebbels, Joseph, 211
Gold, Simone, 42–43
Golden Dome missile defense system, 260
gold standard, 107
Gold Standard Act of 1900, 102
Goldwater, Barry, 242
golf ball: imaginable life of, 331–33
Goodwin, Michael, 117
Google, 131, 299
Google Gemini, 325
Gorbachev, Mikhail, 76
Grand Canyon mid-air collision (1956), 291–92
Grant, Ulysses S., 97
The Grapes of Wrath (film), 175, 176, 176
Graphic Processing Unit (GPU), 302, 302–3, 304, 305
Great Barrington Declaration, 64–65, 66–67
Great Depression, 71, 147–48, 175, 176–78, 263
Greenland: China's interest in, 194–95; Danish control over, 196; early cultures in, 196; European settlers in, 196; fishing in, 203; icebergs in, 204; iconic view of, 197; international airports project, 194–95; Inuit population of, 196; Lutheran church in, 200–201; map of, 194; mosquitos in, 203; strategic location of, 197; Trump's interest in, 194; US offers to buy, 195
Greenland National Gallery of Art, 205
Greenstein, Fred: The Hidden Hand Presidency, 32
Grok (AI assistant), 325; generated response to "No Kings" protests, 279, 281–82
Grove, Andrew, 299
Growth Share Matrix, 231–32, 232
Gruner & Company, 179
Guiliani, Rudy, 316, 323
Gupta, Sunetra, 64, 65

## H

H-1B work visas, 295
Hamilton, Alexander, 173, 174, 174–75, 224
Hanna, Mark, 100, 101, 104
Hannity, Sean, 22, 125, 321, 323
Hansen, Aviaja Rohmann, 200
Harbor Freight, 234
Harding, Florence, 58
Harding, Warren G.: "America First" campaign, 52–53, 53, 55, 58, 297; anti-lynching position, 54–55; daughter of, 58; death of, 57; photo of, 58; presidency, 53–54, 57–58; on racial equality, 54; tariff policy, 107
Harley-Davidson, 233
Harrison, Benjamin, 97
Hawaii: COVID management in, 66; schools in, 82; US annexation of, 102
Hayes, Rutherford B., 97–98, 99
healthcare system, 143, 144, 259, 266, 270
Hearst, William Randolph, 102
Hemingway, Ernest, 139; A Farewell to Arms, 31; The Sun Also Rises, 139, 263
Henry, Patrick, 60, 222–23, 224, 225
herd immunity, 48–49, 64

Heritage Foundation, 140–41
Hernandez, Edgar, 13, 14
The Hill, 35
Hillsdale College, 230
Hinton, Geoffrey, 303
Hitler, Adolf, 75, 76
"hoaxes," 118
Homestead Strike of 1892, 101
Hong Kong: COVID-19 pandemic, 19; flu of 1968, 23–24
Hoover, Herbert, 55–57, 107, 122, 149; American Individualism, 55
Hoover, Herbert Clark, 206, 207
Hoover Institution, 35
Hortman, Melissa, 246
House Committee on Un-American Activities (HUAC), 31–32
House Ways and Means Committee, 99
Howard Johnson's hotel brand, 233
Huang, Jensen, 299, 303
Huffpost, 115, 122–23
Hughes, Howard, 178
Hughes Tool, 178–79
Huntley, Chet, 277
Hussein, Saddam, 164
hydroxychloroquine, 43–44

# I

Ibarra, Jose Antonio, 315
IBM, 186–87, 188
Iceland, 196, 270
Illinois: congressional districts, 225
Ilulissat, Greenland, 194
immigration: Biden's policies on, 171; crimes and, 312, 314–16, 315; economic advantages of, 297–98; illegal, 28–29, 311, 314; restrictions on, 54, 300; technological progress and, 298–99; Trump's policies on, 262, 297
Immigration Act of 1924, 296–97, 300
India: defense spending, 268; STEM graduates in, 298, 299
Indiana: gross state product, 73
India Today website, 43
Inflation Reduction Act (IRA), 170
influenza pandemics: in Asia, 23; conspiracy theories about, 39–40; death toll from, 14, 36; preparedness strategy, 10–11, 12–13; in Russia, 39; in the United States, 5–7, 10–12, 23–24; vaccination and, 11–12, 14; viruses causing, 5–6, 13–14
Infowars, 22, 33, 322
Ingraham, Laura, 230, 295, 323
Inhofe, Jim, 293
Intel Corporation, 299
Interstate Highway System, 29–30
Iowa: COVID-19 pandemic in, 33–34, 65–66; population statistics, 65; schools in, 82
iPhone, 275
Iran: American hostage crisis, 163–64; fanatical theocracy in, 166; Islam faction in, 164; Mosaddegh's government, 160–62; oil industry of, 160; revolution in, 163; Shah's rule in, 162–63; US relations with, 161–62
Iran Air 655 downing, 165
Iran-Contra scandal, 310
Iraq-Iran War, 164–65
ISIS, 308
Italy: COVID-19 pandemic, 20
Ivanovsky, Dmitri, 3–4, 4, 5
ivermectin, 43–44

# J

Jackson, Jimmie Lee, 244
January 6 Capitol riot, 170, 311, 320
Japan: attack on Pearl Harbor, 55, 288; cars exports, 90;

COVID-19 pandemic, 19; defense spending, 268; homicide rate in, 313; US support of, 76
Jefferson, Thomas, 173, 174, 174, 175, 250
Jenner, Edward, 61
Jennings, Ken, 188
Jeopardy, 188
John Deere, 234
Johns Hopkins Center for Health Security, 42
Johnson, Lyndon B., 243, 245
Johnson, Mike, 281
Jones, Alex, 22, 33, 111

## K

Kalanick, Travis, 130, 131
Kangerlussuaq, Greenland, 193
Kapor, Mitch, 131
Karaganov, Sergei, 78
Kasparov, Garry, 186, 187
Kennan, George F., 76–78
Kennedy, John F., 32, 242–43
Kennedy, Robert F., 245
Kennedy, Robert F., Jr., 41, 42, 62, 110, 236–37, 322
Kentucky Fried Chicken restaurant chain, 231, 233
Khomeini, Ruhollah, 163
Khosrowshahi, Dara, 131
King, Martin Luther, Jr., 242, 243, 244–45
Kirk, Charlie: on abortion, 251; assassination of, 242, 245, 246, 253, 307; on DEI policies, 252; education of, 247; on gun control, 251; on LGBTQ, 251–52; photos of, 243, 248; political activism, 247–48, 249; popularity of, 248; social media broadcasts, 249, 250, 252; support for Trump, 248, 249; support of "Seven Mountains Mandate," 251; vision of Christian government, 250–51; on women's role in family, 252

Kirk, Erika, 253
Kistiakowsky, George, 299
Kodak Company, 233
Korean War, 28, 136
Koresh, David, 308–9
Krizhevsky, Alex, 303
Ku Klux Klan, 244
Kulldorff, Martin, 64, 65
Kurlansky, Mark: *Salt: A World History*, 9

## L

Laffer, Arthur, 56
Lakanwal, Rahmanullah, 306
Lamont, Thomas W., 207
Land O Lakes agricultural cooperative, 231
Landon, Alf, 150
LaPierre, Wayne, 310
Laplace, Pierre-Simon, 273
Large Language Model (LLM), 301
Las Vegas shooting (2017), 246
League, Archie, 289, 289–90
League of Nations, 55
LeBlanc, Bibi, 211, 214, 215
LeMay, Curtis, 213
Leo XIII, Pope: *Rerum Novarum* ("Of New Things") encyclical, 126–27, 129, 133
Leo XIV, Pope: education of, 124; election of, 124; MAGA supporters and, 127; missionary in Peru, 124, 125; social activism of, 126; Trump and, 127–28
leprosy, 62
*Let's Make a Deal* (TV show), 271, 275
LGBTQ, 251–52
Limbaugh, Rush, 21, 110
Lincoln, Abraham: AI-generated address to Americans, 325–26, 327–28; assassination of, 243; Gettysburg address, 305; letters to Russian Czar, 10; mourning for, 57
Lindbergh, Charles, 289

Lindell, Mike, 323
Little Rock Nine, 307
Liuzzo, Viola, 244
logical reasoning, 181–82, 183
London Great Plague (1665), 39, 40
London's National Institute for Medical Research, 7
Long, Huey, 148
Longworth, Alice Roosevelt, 111
Loomer, Laura, 127
Lopez, Maria, 282
lynching, 54–55

# M

Maddox, Lester, 243
Madison, James, 175, 222, 223–25, 250
Maine: COVID-19 pandemic in, 66
Make America Great Again (MAGA), 26, 27, 58, 249, 287
Manhattan Project, 298
Marconi, Guglielmo, 299
Marshall, George, 212
Marshall Plan, 212–13
Maryland: COVID-19 pandemic in, 66; schools in, 82
Masino, Julie, 229, 230
Massachusetts: redraw of electoral districts in, 223, 227; schools in, 82
Mattis, Jim, 195
McCarran-Walter Act of 1954, 54
McCarthy, Joseph, 31–32
McDonalds, 233
McEntarfer, Erika, 208, 238
McKinley, Ida (McKinley's daughter), 98
McKinley, Ida (née Saxton), 98–99, 103
McKinley, Katie, 98
McKinley, William: assassination of, 103, 243; campaign poster, 98; children of, 98; Civil War veteran, 97; Gold Standard Act of 1900, 102; legal practice of, 99; marriage of, 98; military service of, 97; monetary policy of, 107, 108; Ohio governorship, 100; political career of, 99, 106, 109; presidential election, 102; speech at Pan-American Exhibition in Buffalo, 106–7; study of law, 97; support for Rutherford B. Hayes, 97–98; tariffs policy, 99–100, 101, 104
McMillan, Doug, 133
McVeigh, Timothy, 309–10
measles, 49, 62
Medicaid, 143, 144, 259, 266
Medicare, 143, 144, 266
Mellon, Andrew, 55–56, 137
MERS-CoV (Middle East Respiratory Syndrome CoronaVirus), 17–18
Mexico: emigration from, 28–29; fiscal crisis, 270; labor shortages in, 28; US aid to, 287
Michigan: "American Patriot Rally," 34; COVID-19 pandemic in, 34
Micro-Particle Colloidal Silver Generator, 22
Microsoft, 299
Microsoft Windows, 232
Milei, Javier, 284, 284–86
Minsky, Marvin, 187, 188–89
Mississippi: schools in, 82
MNIST dataset, 191
Modelo Especial, 233
"Molly Maguires" (secret labor organization of Irish miners), 126
Monroe, James, 224, 225
Montgomery, Richard, 59, 60
Montgomery, William, 248
Monty Hall Problem, 271, 272, 274–75
Moore, Oneal, 244
Moore, Steven, 210
Morgan, J.P., 101

Morgenthau, Henry, Jr., 211
Morrill Tariff Act, 105
Morris, Gouverneur, 222
Mosaddegh, Mohammad, 160–61, 161, 162
Mount Sermitsiaq, 198
Moynihan, Daniel Patrick, 44–45
MSNBC, 121–22
Mulvaney, Dylan, 233
Munich Agreement, 75
Murdoch, Rupert, 323
Murrow, Edward R., 277
Musk, Elon, 90, 91, 146, 264, 284, 285, 299

# N

Nashville, TN: crime statistics, 314, 315; foreign-born population in, 315
National Bureau of Economic Research, 297
National Business Aviation Association (NBAA), 294
national debt, 136–37, 138, 218; accumulation of, 138–39, 221, 260, 263, 264, 269–70; cost of financing, 264–65; as percentage of GDP, 30, 30, 138, 260, 262–63, 270
National Education Association (NEA), 79
National Institute for Medical Research, 7
National Institutes of Health (NIH), 114
National Origins Act of 1924, 54
National Rifle Association (NRA), 310
National States Rights Party, 244
National Technical Honor Society, 307
Nebraska: COVID-19 pandemic in, 33–34, 66; schools in, 82
Nelson, George "Baby Face," 318
New Deal programs, 150, 175, 177
New Hampshire: COVID-19 pandemic in, 65–66; population statistics, 65

New Jersey: Airway Traffic Control Station in, 290; COVID-19 pandemic in, 66
New Mexico: COVID-19 pandemic in, 66; schools in, 82
news media: artificial intelligence and, 277, 278–79; Fairness Doctrine and, 277–78; level of trust in, 277–78; misinformation in, 278, 322, 323; political polarization and, 278; reports on nation's economy, 239–40
Newsom, Gavin, 222
New York City: CompStat system in, 317; crime statistics, 314, 315, 316, 317; foreign-born population in, 314–15, 318; police force in, 317
New York Daily News, 242
The New York Herald, 39
New York Post, 117–18, 252
New York State: COVID-19 pandemic in, 66
The New York Times, 36, 37, 115, 119–20, 273; on Trump's first 100 days in office, 119–20
Nike Corporation, 234
9/11 Terrorist Attacks, 12, 157, 165, 268, 269
Nixon, Richard, 27, 32, 47, 245
No Child Left Behind Act (NCLB), 80–81, 83
Noem, Kristi, 51
"No Kings" protests, 279–82
Norquist, Glenn, 216, 217
North American Treaty Organization (NATO), 73–74, 76–78, 169, 171
North Carolina: redraw of electoral districts in, 226
North Dakota: COVID-19 pandemic in, 33–34; schools in, 82
Nuuk, Greenland, 193, 198
Nuuk Fiord, 203, 204

Nuuk Golf Club, 202
Nvidia, 299, 301–3, 302, 305

## O

Oath-Keepers, 311
Obama, Barack, 113, 114, 157, 158, 208
Office of Strategic Services (OSS), 162
O'Handley, Rogan (aka DC Draino), 234
Ohio: industrialization of, 106
Ohio Gang, 57, 58
Oil Embargo of 1973, 89
oil industry, 9, 90, 120, 160, 161, 162, 178–79
Omnibus Budget Reconciliation Act, 140, 217
One Big Beautiful Bill, 259, 260, 267
OpenAI Company, 303
Operation Ajax, 162
Operation Allies Welcome, 306
Operation Warp Speed, 68, 170
Operation Wetback, 28
optical character recognition, 189–91, 190
Ordo Amoris ("order of love"), 125
Oregon: schools in, 82

## P

Pan-American Exhibition in Buffalo, New York, 103, 106
Pandemic and All-Hazards Preparedness Act (Preparedness Act), 12–13
Panic of 1893, 101
Pasteur, Louis, 3
Patel, Kash, 320
Pearl Harbor attack, 55, 288
Pearl Milling Company, 231, 301
Pence, Mike, 37
Pennsylvania: mining strike, 126; redraw of electoral districts in, 225–26; Whiskey Rebellion, 105
Perdue, Frank, 86
Peron, Juan, 284
Peru: forced sterilization of indigenous women, 124
Peterson, Jordan, 247
Philippine Islands, 102
Philippines: COVID-19 pandemic in, 19
Pirro, Jeanine, 323
Plandemic: The Hidden Agenda Behind Covid-19 (video), 41
Plumbers and Steamfitter's Union, 26
Poland: NATO membership, 76
polio vaccination, 29, 62–63
Powell, Jerome, 238
Powell, Sydney, 323
Prevost, Robert. See Leo XIV, Pope
Pritzker, Jay Robert, 282
Programme for International Student Assessment (PISA), 84
Project 2025, 141
Protection Pledge, 217–18
Proud Boys, 311
Public Health, and the Terrifying Bioweapons Arms Race (Kennedy), 237
Pulitzer, Joseph, 115
Pullman Strike of 1894, 101
Pulse nightclub shooting, 246

## Q

Qaqortoq, Greenland, 194
Qooqqut Nuan summer camp, 203
Quinnipiac University survey, 48

## R

radio, 299
Radio Act of 1927, 56
Radio Corporation of America (RCA), 299
Radzinsky, Edvard: Alexander II: The Last Great Tsar, 9
Ramaswamy, Vivek, 83–84
Rand, Ayn, 175; Atlas Shrugged, 57

Rao, Neomi, 169
Reagan, Ronald: AI-generated address to Americans, 325–26, 329–30; Cold War diplomacy, 78; economic policy of, 141, 217; education policy of, 80; Fairness Doctrine and, 277; on Great Depression, 147; Gulf War and, 165; Iran hostage crisis and, 164; "Let's Make America Great Again" campaign slogan, 27; media policy of, 277; national debt under, 141; The Pledge, 216; Social Security policy, 151, 261, 262, 267; tariff policy of, 90; tax policy of, 137; on Vietnam War, 241
The Real Anthony Fauci (Kennedy), 237
Reeb, James, 244
Republican Party: air traffic control system and, 293; efforts to reduce federal deficit, 259–60; "No Kings" protests and, 279–80, 282; position on economic policies, 90, 100, 102, 104, 122; position on Social Security, 150; redrawing of election districts, 226; support of Great Barrington Declaration, 67–68; Trump's impact on, 117, 155, 172, 262; vaccine hesitancy and, 69, 70
Republican Pledge, 216
Rerum Novarum ("Of New Things") encyclical, 126–27, 133
Revenue Act of 1921, 56
Riley, Laken, 315
Robb Elementary School shooting (2022), 246
Robin Hood, 31
Robinson, Tyler James, 253, 307–8
Rockefeller, John, 100–101
Roosevelt, Franklin D.: 1933 inaugural address, 177; 1936 presidential election, 177–78; AI-generated address to Americans, 325–26, 328–29; New Deal programs, 123, 150, 175, 177; Social Security policy of, 149–50; at Tehran Conference, 162
Roosevelt, Theodore, 102, 111
Routh, Ryan Wesley, 307
Roy, Chip, 12
Royal Greenland fishing trawler, 199
Rucho v. Common Cause, 226
Rumsfeld, Donald, 241
Russell, Kevin, 254
Russian Federation: defense spending of, 77, 268; economy of, 73; fiscal crisis in, 270; NATO and, 73, 76, 78; STEM graduates in, 298, 299
Russian flu pandemic of 1889, 39
Russo-Ukrainian war, 153, 171
Rutter, Brad, 188

## S

Saint Lawrence Seaway, 179
salt: as strategic material, 9
Sanders, Bernie, 280
Sandy Hook Elementary School shooting (2012), 246, 322
San Francisco, CA: crime statistics in, 314, 315; foreign-born population in, 314–15, 318
SARS-CoV-1 (Severe Acute Respiratory Syndrome CoronaVirus), 15–16, 18
SARS-CoV-2 (Severe Acute Respiratory Syndrome Coronavirus 2). See COVID-19 pandemic
Saudi Arabia: defense spending in, 268; MERS outbreak in, 17–18; oil industry of, 160; Trump's ambassador to, 239

Savings and Loan Crisis of 1987, 263
Scalise, Steve, 246
Scavino, Daniel, 47, 47–48
Schiffmann, Avi, 19, 20
Schlitz beer, 234
Sean Hannity Show, 321, 322
The Searchers (film), 57
Selma Freedom Marches, 244
"Semper Fidelis" (song), 256
"Seven Mountains Mandate," 251
Seventh Military World Games, 112, 113
Shannon, Claude, 182
"Share Our Wealth" proposal, 144
Sharia law, 251
Shore, Dinah, 134
Siegel, Mark, 21
Silver Edge company, 22
Simonyi, Charles, 299
Sinclair, Upton, 148
slavery, 10
smallpox: outbreaks of, 59–60; vaccinations from, 60–61, 62
Smoot-Hawley tariffs of 1930, 206, 207
social media: campaign against Dr. Fauci on, 47, 47–48; conspiracy theories and, 41; political divide and, 246
Social Security Act of 1935, 150
Social Security programs: crisis of, 144, 150–51, 262; criticism of, 145–46; federal deficit and, 264, 270; Full Retirement Age and, 267; origins of, 145, 146–48; taxation and, 143, 143, 144, 150, 259, 260–62, 267
Sousa, John Phillip, 256
Sousa Band, 256
South Dakota: COVID-19 pandemic in, 33–34, 66; schools in, 82
Southeast Asia: consumption of wild animals in, 17
Southern Fried Chicken, 233

South Korea: COVID-19 pandemic in, 20; fiscal crisis in, 270
SpaceX, 299
Spanish-American War, 102
Spanish Flu of 1918 (H1N1 influenza): at Camp Funston, Kansas, 6; death toll, 23, 52; spread of, 5–8, 9, 10
Stalin, Josef, 74–75, 162
Stanford Medical School, 49
Starkloff, Max, 7
"Stars and Stripes Forever" (song), 256
Stasi, 211
Steak and Ale restaurant chain, 233
Steak 'n Shake fast food chain, 231
Steinbeck, John, 175
Steinmetz, Charles, 298
Steve Deace Show, 48–49
St. Louis Lambert Airport, 289
Stock Market Crash of 1928, 263
stolen election claim, 311, 322–24
strikes, 101, 126
Sturgis Motorcycle Rally, 49–51
Substack posts, 331
Sutton, Willie: That's Where the Money Was, 319
Swalwell, Eric, 118
Sweden: COVID deaths in, 38; education system in, 84
Swine Flu Pandemic of 2009, 13–14, 36
Swope, Gerard, 146, 148–49
Szilard, Leo, 299

## T

Taber, John, 145
Taiwan: COVID-19 pandemic in, 20
Taliban, 308
Target Corporation, 234
Tariff Act of 1890, 100, 104
tariffs: annual revenue from, 220, 221; on Chinese goods,

220; Hoover's policy on, 206–7; impact on profits, 218–19; McKinley's end of, 107; Reagan's policy on, 90; reciprocal, 94–95; Trump's policy on, 94–95, 169

Tax Cuts and Jobs Act (2017), 169

taxes: avoidance of, 55–56, 138; budget deficit and, 140; Clinton's policy on, 141–42, 143–44; impact on business, 137–38; Mellon's policy on, 55–56; revenues from, 142–43; Trump's policy, 138, 144, 169, 260

Tea Party, 249

Tebb, William: Leprosy and Vaccination, 62

Tehran Conference (1943), 162

Teller, Edward, 299

Tesla (company), 232, 299

Tesla, Nikola, 39, 298

Tesla Model Y, 90–91

Texas: COVID-19 pandemic in, 68–70, 69; health care system in, 68–69; protests against lockdown, 33; reorganization of electoral districts, 222, 228

Texas State Legislature, 222

Thomas, John, 59

Three-Percenters, 311

Thule People, 196

tobacco plant disease, 3–4

Toronto Star, 48

Townsend, Francis, 148

Townsend Clubs, 148

Toyota Corolla, 89

Toys "R" Us, 233

Tree of Life Synagogue shooting (2018), 246

Truman, Harry S., 78, 161, 213

Trump, Donald: assassination attempts on, 246, 307; as Biblical Samson, 248, 249; on birthright citizenship, 295–96; ChatGPT analysis of presidency of, 167–70, 171–72; ChatGPT psychological profile of, 152, 154–55; communication style of, 154; congratulations to Cardinal Robert Francis Prevost, 127; COVID-19 management, 20, 20–21, 23, 38, 46, 48, 64, 71–72, 170; crack down of violent crimes, 312–13, 314; Deep State foes of, 248, 285, 311; Department of War, 240–41; deregulation efforts, 113–14; disparagement of the BLS, 208, 209; economic policy of, 169, 171; education policy of, 79, 80, 84; employment statistics under, 238; executive orders of, 120; Fauci and, 46–48; federal government shutdowns and, 259; financial aid for Argentina, 283, 286–87; firing of Erika McEntarfer, 208, 210; first 100 days in office, 115–23; foreign policy of, 169, 171; on H-1B work visas, 295; immigration policy of, 295, 297, 300, 306–7; interest in Greenland, 194, 195; leadership style of, 153–54, 157–58, 172; MAGA agenda, 26, 27, 58, 249, 287; on making America great again, 117; on "No Kings" protests, 282; official portrait of, 153; One Big Beautiful Bill, 259, 260, 267; populism of, 159, 168–69, 172; proclivity for alternative facts, 45; on redraw of Congressional districts, 228; second inaugural address of, 104; slash of humanitarian aid, 287; social media posts on Cracker Barrel, 230, 231; supporters of, 158; tariff

policy of, 94–95, 107, 169, 206, 209, 218, 220, 240; taxation policy of, 138, 144, 169, 260, 262; Vigano's letter to, 41–42
The Truth about Covid-19 (Kennedy), 236
Turing, Alan, 186, 188, 273, 275–76
Turkey: fiscal crisis in, 270
Turning Point USA (TPUSA), 248, 249
2008 Financial Crisis, 263, 269
2025 Potomac River mid-air collision, 288

## U

Uber, 129, 130–32
Ukraine: defense spending, 268
Ulam, Stanislaw, 299
Uncle Ben's Rice, 231
unemployment, 101, 147, 149, 169, 209, 240
Union of Soviet Socialist Republics (USSR): collapse of, 73, 76, 78; US relations with, 161–62; World War II and, 74–76
United Kingdom: defense spending, 268; education system in, 84; GDP, 106; National Air Traffic Services in, 293
United States: in the 1950s, 134–35; banned books in, 31; Christian theology as foundation of, 242, 250–51; communism scare in, 27, 31–32; COVID-19 pandemic in, 20–23, 24–25; defence spendings, 260, 262, 267–68, 268; division in, 93–94, 142, 173; economic development of, 138; education system of, 79–82, 83–85, 85; employment statistics, 207, 208, 209; flu pandemics in, 5–7, 10–12, 23–24; foreign-born population in, 313–14, 315; GDP, 96, 106, 141, 142–43; government distrust in, 310–11; healthcare in, 266; homicide rate in, 313, 317; income gap, 135–36; investment in, 116; Iraq-Iran War and, 164–65; isolationism of, 55; mass killings in, 246, 309–10, 322; patriotic mood in, 173; political polarization in, 157–58, 159, 174, 226–28, 278; post-war policy toward Germany, 211–15; STEM graduates in, 298, 299; stereotypical depiction of, 135; trade imbalances, 94–95, 96, 206, 219; transportation system, 29–30; violence in, 103, 170, 243–46, 307–8, 314–18, 315; workforce, 94
United States Agency for International Development (USAID), 119
United States Citizenship and Immigration Services (USCIS), 306
United States Virgin Islands, 195
US Congress: holding funding bills, 258–59; responsibility for federal deficit, 263, 269, 270; salaries of members of, 258; shutdowns of federal government and, 258, 259, 259, 294
US Constitution: on election principles, 222, 223, 224, 226; guarantees of individual freedoms, 224, 250; "No Kings" Protests and, 279; on power of states, 83
US Marine Band, 256
USS Nimitz, 163
US Supreme Court: interpretation of 14th

Amendment, 296; ruling on gerrymandering, 226
USS Vincennes, 165
Utah: COVID-19 pandemic in, 33–34, 66; schools in, 82

## V

vaccination: conspiracy theories about, 62; risks of, 61, 62–63; scepticism about, 61–62, 69
Vaccination Act of 1853, 61
Vaccination Act of 1898, 61–62
vaccine manufacturing, 11–12
Vance, James David, 124–25, 295–96, 296, 297, 299
Vance, JD, 124–25
Vanderbilt, Cornelius, 101
Vax-UnVax (Kennedy), 237
Vermont: COVID-19 pandemic in, 68
Vermont State University, 254–55
Veterans Administration (VA), 266
Victoria Station, 233
video games, 302, 303, 304
Vietnam: trade imbalance with the US, 95–96
Vietnam War, 241, 310
Vigano, Carlo Maria, 41–42
Virginia: Christian teaches in, 250; national elections in 1789, 223; schools in, 82
viruses, 3–6, 7–8
Von Neumann, John, 299

## W

Wakefield, Andrew, 62
Wallace, George, 243
Wall Street Journal, 321
The Wall Street Journal, 37, 115, 119, 264, 265
Walmart Corporation, 132, 133, 218
Warsaw Pact, 74, 76
Washington, George, 28, 60, 105, 173–74, 230, 240, 255, 325–27
"Washington Post March" (song), 256
Washington State: COVID-19 pandemic in, 66; schools in, 82
Watergate scandal, 27, 310
Wayne, John, 57, 134
Weaver, Randy, 308
Weisskopf, Victor, 299
Welch, Jack, 93
Welles, Orson, 31
Welsh, Joseph, 32
West Berlin, 211, 213, 215
We The Presidents (Gruner), 261
wet markets in China, 113
whiskey tax, 104–5
White Tower, 233
Whitman, Walt: Leaves of Grass, 31
Wigner, Eugene, 299
Williams, Denise Halvorsen, 214
Willis, Mikki, 41
Wilson, Charlie, 26
Wilson, Woodrow, 53, 55
Wilson-Gorman Tariff Act of 1894, 100
Wisconsin: COVID-19 pandemic in, 66
The Wizard of Oz (film): as political allegory, 105, 108–9
Wolfe, Andrew, 306
Wolfenstein (video game), 302
women: activism, 281; family responsibilities, 252; forced sterilization of, 124; reproductive health of, 171, 251, 258; in sports, 116; voting rights, 53
World Economic Forum, 42
World Health Organization (WHO): conspiracy theory about, 41; COVID-19 and, 19, 20, 23; MERS and, 18; SARS and, 16; Swine Flu pandemic and, 14
World War II, 75
Wuhan Institute of Virology, 16,

111, 112
Wyoming: COVID-19 pandemic in, 33–34; schools in, 82

## X

X (social network), 230, 282
X, Malcolm, 244

## Y

yellow fever, 5
Yeltsin, Boris, 74
Younge, Samuel, Jr., 244
Youngkin, Glenn, 281

## Z

Zeitz, Jochen, 233–34, 235

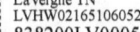
www.ingramcontent.com/pod-product-compliance
Lightning Source LLC
LaVergne TN
LVHW021651060526
838200LV00050B/2299